Henry James Clark

Mind in nature; or The origin of life, and the mode of development

of animals

Henry James Clark

**Mind in nature; or The origin of life, and the mode of development of animals**

ISBN/EAN: 9783337229412

Printed in Europe, USA, Canada, Australia, Japan

Cover: Foto ©berggeist007 / pixelio.de

More available books at **www.hansebooks.com**

# MIND IN NATURE;

OR

## THE ORIGIN OF LIFE, AND THE MODE OF

## DEVELOPMENT OF ANIMALS.

BY

### HENRY JAMES CLARK, A.B., B.S.

ADJUNCT PROFESSOR OF ZOÖLOGY IN HARVARD UNIVERSITY,
CAMBRIDGE, MASS.

MEMBER OF THE AMERICAN ACADEMY OF ARTS AND SCIENCES, BOSTON, MASS.; OF THE BOSTON SOCIETY OF NATURAL HISTORY; CORRESPONDING MEMBER OF THE AMERICAN MICROSCOPICAL SOCIETY OF NEW YORK, ETC., ETC.

"La naissance des êtres organisés est donc le plus grand mystère de l'économie organique et de toute la nature; jusqu'à présent nous les voyons se développer, mais jamais se former." CUVIER, *Règne Animal*, 1829.

*WITH OVER TWO HUNDRED ILLUSTRATIONS.*

NEW YORK:
D. APPLETON AND COMPANY,
443 AND 445 BROADWAY.
1865.

# PREFACE.

THE following chapters comprise the substance of a course of public lectures which were delivered by the author in the hall of the Lowell Institute, in Boston, Mass., during the months of February and March, 1864. Since that time a considerable amount of matter has been added, chiefly in the form of notes; and some alterations have been made in the sequence of the subjects, during which it has been found necessary to arrange the chapters in such a way that they do not correspond with the original succession of the lectures. To obviate any difficulty, however, which might arise from the frequent references to previous lectures, care has been taken to mention the page upon which the subject alluded to may be found.

Although these lectures were originally given to the public in a popular form, it must not be supposed that they were altogether based upon what was already known to the scientific world; on the contrary, no small proportion of the facts and ideas promulgated herein are claimed by the author to be original with himself. In this respect the attention of naturalists is invited to the following subjects. The structure of *Bacterium termo*. The organization of *Vibrio baccillus*. The theory of the egg. The polarity and bilaterality of the egg. The cellular structure of *Actinophrys*. The relation of vitality to the various degrees of organic complication. The relation of the egg to secondary causes. Anatomy of *Polypi*. The nervous system of *Infusoria*. The individuality of *Hydro-Medusæ, Strobiloid Medusæ*, and the lower Invertebrata. The peculiar mode of reproduction of the hydroid genus *Rhizogeton. Monstrosity* and reproduction by budding. *The typical animal.* The relation of the five grand divisions to the typical animal. The typical form of *Protozoa* and *Zoöphyta.* Polarity of animals. The so-called radiation of *Zoöphyta.* A new form of ciliated Infusoria, *Heteromastix.* The digestive system of Infusoria. The longitudinal axis of Echinodermata. Anatomy of the apodous holothurian, *Caudina.* Investigation of the structure of *Loligopsis*, in reference to the dorsal and ventral sides of *Cephalopoda.* The relation of the spinal marrow to the vertebral axis. The development of the tentacles in *Tubularia.* The *bilaterality* of Tubularia. The structure of the *scyphostoma* of Aurelia. The *subsidiary layer* of the vertebrate embryo. In this connection the author would also request, especially of the younger naturalists, the perusal of the "*Note on Scientific Property,*" on p. 37, and also the note on p. 304.

A large proportion of the figures which were used to illustrate the lectures are reproduced in this volume; and, beside these, numerous others are introduced in order to meet the requirements of the additional matter and the notes. At least two thirds of these were taken from the original drawings of the author, and the rest were almost invariably selected from the first authorities. In the latter case, the name of the original author is given at the end of the description of the figure.

Whenever possible, figures of American animals were selected in preference to those from foreign sources.

When an object is magnified, the amount of enlargement is stated to be so many *diameters*, e. g. 250 diam. i. e. 250 times the diameter of the natural size.

HENRY JAMES CLARK.

CAMBRIDGE, Mass., *September*, 1865.

# CONTENTS.

## PART FIRST.

### THE ORIGIN OF LIFE.

#### CHAPTER I.

THE OLD AND THE NEW DEVELOPMENT THEORY. — CHEMICAL AFFINITY AND THE PRINCIPLE OF LIFE, OR THE ORGANIC AND INORGANIC FORCES. — THE LOWEST ORGANIC BEINGS. — SPONTANEOUS GENERATION . . 3

#### CHAPTER II.

WHAT SPONTANEOUS GENERATION PROVES. — THE EGG IS THE LOWEST PHASE OF ANIMAL LIFE. — THE EGG A BIPOLAR ANIMAL. — THE EGG CONTRASTED WITH THE LOWEST FORMS OF ADULT BEINGS. — THE EGG AS RELATED TO SECONDARY CAUSES . . . . . . . 28

#### CHAPTER III.

THE OLD APHORISM, "OMNE VIVUM EX OVO," NOT STRICTLY CORRECT. — THE ORIGIN OF INDIVIDUALS BY BUDDING AND SELF-DIVISION . . 54

#### CHAPTER IV.

THE REGENERATION OF LIVING ORGANISMS AFTER PARTIAL DESTRUCTION. — THE PERSISTENCY OF VITALITY DURING DECOMPOSITION. — THE RELATION OF SECONDARY CAUSES TO THE GREAT PRIMARY CAUSE. — ANIMALS PRIMARILY CREATED IN AN ADULT STATE . . . . . . 88

#### CHAPTER V.

SPONTANEOUS GENERATION AND REPRODUCTION BY BUDDING AND FISSIGEMMATION MOST FREQUENT AMONG THE LOWEST RANKS OF ANIMATE BEINGS. — ALL ANIMALS ALIKE IN THE EARLIEST STAGES. — MAN AND MONAD ARE AT ONE TIME A MERE DROP OF FLUID . 109

## PART SECOND.

### THE FIVE GREAT ANIMAL GROUPS.

#### CHAPTER VI.
THE IDEAL TYPES. — ALL ANIMALS BILATERAL . . . . . 117

#### CHAPTER VII.
THE DISTINCTION BETWEEN ANIMALS AND PLANTS. — THE PSEUDO-INFUSORIA, THEIR PLANT-NATURE. — THE PLANT-LIKE INFUSORIA . . . 131

#### CHAPTER VIII.
THE PHYTOZOA, OR PLANT-ANIMALS. — THEIR RELATION TO UNDOUBTED ANIMALS . . . . . . . . . . . . . 153

#### CHAPTER IX.
THE SYMBOLICAL ANIMAL. — THE PROTOZOA . . . . . . . 158

#### CHAPTER X.
ZOÖPHYTA . . . . . . . . . . . . . . 177

#### CHAPTER XI.
MOLLUSCA . . . . . . . . . . . . . . 195

#### CHAPTER XII.
ARTICULATA . . . . . . . . . . . . 214

#### CHAPTER XIII.
VERTEBRATA . . . . . . . . . . . 226

#### CHAPTER XIV.
THE DISTINCTNESS OF THE FIVE GRAND DIVISIONS . . . . . 236

#### CHAPTER XV.
THE MIMETIC FORMS OF DIVERSE TYPES OF ANIMALS . . . . 248

#### CHAPTER XVI.
THE TRANSITIONS AMONG THE SUBORDINATE TYPES OF THE FIVE GRAND DIVISIONS . . . . . . . . . . . 254

## PART THIRD.

### THE MODE OF DEVELOPMENT OF ANIMALS CORRESPONDS WITH THE TYPE OF THE GRAND DIVISION TO WHICH EACH ONE SEVERALLY BELONGS.

#### CHAPTER XVII.

THE SEGMENTATION OF THE EGG. — THE EMBRYOLOGY OF PROTOZOA, ZOÖPHYTA, MOLLUSCA, ARTICULATA, AND VERTEBRATA . . . 283

INDEX . . . . . . . . . . . 317

# PART FIRST.

## THE ORIGIN OF LIFE.

# PART FIRST.
## THE ORIGIN OF LIFE.

### CHAPTER I.

THE OLD AND THE NEW DEVELOPMENT THEORY. — CHEMICAL AFFINITY AND THE PRINCIPLE OF LIFE, OR THE ORGANIC AND INORGANIC FORCES. — THE LOWEST ORGANIC BEINGS. — SPONTANEOUS GENERATION.

No doubt most of you have heard of what is called the "Development Theory," and have been led, in the course of your various readings, to believe that its doctrines have such a tendency as might induce one finally to exclude from the mind every idea of the interposing hand of the Creator, in the origination, development, and growth of living creatures.

It is true that this is one of the various forms of the development theory; and the one which, more than any other, is prominent in the minds of men at large throughout the world.

It is that form of the development theory which teaches that all things originated through physical forces, which operate according to what are called *physical laws;* the laws of electricity, magnetism, chemical affinity, &c.; laws which have been said to *administer themselves.*

There was a time when this idea may be said to have raged among the philosophic portions of the community; but although the heat of the fever has abated at this time, still there is a certain tendency to relapse, if not into the old stage, at least into another no less dangerous form of it, which is this.

Admitting, say the advocates of this theory, that *in the beginning* the Creator made all things living by a *direct act* of his own hand, yet after that, in order that the universe might go on in the course which he had set it upon, he established certain laws which should act as *independent rulers* of the forces of nature, competent to influence the *origin*, the development, and growth of all things, without any further interposition of his power.

This is indeed the most insidious form in which Materialism has attempted to make its first approaches to the citadel of our belief in a ruling Providence. It acknowledges the *primary interposition* of the Great First Cause, but eventually would make it seem possible that his power can be made to operate in such a way that his very existence is not necessary; for, say they, why should he continue to exist, if the laws which he has established can control the work for him? And now then, so much being allowed, the next step in the fallacy is in this guise.

It being admitted that these forces are competent to perform all the offices of a Creator; they being equally powerful to create and to destroy, to determine the movements of this mighty universe or to originate new ones; all this being admitted, why, then, are they not equally powerful and omnipresent as a Creator; what is there not in them that a Creator possesses; if nothing, then why may they not have given rise to their own existence, or have existed eternally; in fine, what more is needed to constitute them a Great First Cause?

Your ready answer no doubt will be that the revelation declares the existence of a God! Truly! And it is this revelation which satisfies a large portion of the community; but then there is still a great body of readers and thinkers, inquiring minds, who would like to know more about the *manner* in which the Creator manifests himself. What did the King of Israel mean, they ask, when he said, " The heavens declare the glory of God; and the firmament showeth his handy work?"

I hope in the course of these lectures to be able to answer this question to your fullest satisfaction. In doing this, it is my design to proceed in an argument to prove that there is a power at work in the universe which possesses foreknowledge; the design of a forecasting, foreordaining mind, — a thinking, intelligent, *animate being;* such a combination of powers that no form of physical law could possibly be conceived to represent. He must be a bold metaphysician indeed, who could assert such a possibility! It seems to me that it would be tantamount to declaring an utter impossibility to be possible.

The form of the argument which I wish to introduce for your consideration is identical, up to a certain point, with just such a one as would be advanced to prove the prevalence of independent physical law as a controlling power; but beyond that point I hope I shall be able to show that it may be used as evidence of a *thoughtful design* to produce a *succession* of events, or a combination of contemporaneous, interdependent phenomena.*

I have chosen for the subject of this course of lectures a somewhat comprehensive field, namely, that of the origin of life and the mode of development of animals, because I do not wish to be limited in what I have to say to a simple narration of the mode of putting together the organization of animals. I have purposely used the term " putting together " here, because that is the general idea of the way in which an organized being is brought into existence. You have been led to think so by various means. In " Paley's Philosophy," a work so extensively

---

* In the succession of beings from a lower to a higher type, and a consentaneous greater degree of complication, we have the strongest proof of an intelligent being, designing, ordaining, and controlling. The laws of the older physicists were not claimed to be derived from an intelligence; they were deemed to exhibit the necessary operations of matter upon matter; but when we see that these laws have an order, and, as they are understood at the present day, a *rate of succession* in their operations, which have the stamp of thoughtfulness impressed upon them, it is impossible not to discover that they do not work of their own accord, but are controlled by a creative forethought and design. If the product of these causes was a heterogeneous mixture of beings, with no relation whatever among themselves, then one might more plausibly claim that the so-called physical causes had produced living creatures. As it is, though, we have before us animals allied to each other by *progressive* relations, which finally, if we follow them up, end in the highest forms of life at the present day, from having begun with the lowest and ascended. What mere non-intelligent causation could produce the like?

used in our schools and colleges, the animal frame is compared to a watch with all its interdependent wheels and pulleys, which can be put together after all the parts have been manufactured separately. In our institutes of instruction we have very little or nothing of the physiology of growth and development taught, because the text-books are devoted to an enumeration of the organs of the adult body; and as the movements of the animal frame are illustrated by mechanical contrivances, pulleys, levers, and the like, the digestion by chemical forces, and the circulation by mechanical propulsion, it is very natural that scholars should grow up with the idea that the human frame is moulded upon a mechanical contrivance.

This is of what I would totally disabuse your minds. Although there may be a certain degree of truth in it, yet it is in such a small proportion to what is commonly received to be the truth, that I would rather, for the present purposes, you had never heard of such a thing as an organized being, for then your minds would not be diverted, by any preconceived ideas, from the argument which I am about to lay before you. I beg, therefore, that you will allow me to lead you on, unrestrainedly, step by step, in the new path which I have laid out for the present occasion.

In the preparation of these lectures I have revised the whole history of the *origin of life, and of the mode of development of animals*, as it is understood at the present day.

I have commenced with the lowest and simplest forms of life; those *obscure manifestations of a living existence*, immediately upon, or rather just this side of the confines of mere *chemical* association. The characteristics of these I have carefully balanced between the probabilities of *life* on the one hand, and of *mere existence without life* on the other.

This no doubt seems to you like a mystery; and so it is, in a measure at least; and I would, certainly, rather that it might so appear at the outset, than that it should be involved in your minds with any of the mechanical ideas of which I have spoken, when referring to Paley's work.

All *living beings*, whether animals or plants, are composed, essentially, of four chemical elements, Carbon, Hydrogen, Oxygen, and Nitrogen, which are combined in various proportions.

It would be very natural to suppose after this statement that these chemical combinations are such as you see exhibited every day around you; such as are called the *natural* chemical affinities, which exist, for instance, between the gases of which the air is composed, or the acids, chloroforms, alcohol, salts, crystals, &c., &c. But this is not so! Although I am well aware that it is getting to be the general opinion, among organic chemists and physiologists, that the inorganic and organic affinities approximate each other, and may eventually turn out to be, among themselves, mere degrees of difference; yet, even with such an idea in view, it is not incorrect to say that it is in direct opposition to natural chemical affinity that organized beings exist.

There is another principle or affinity which is not commonly recognized in our daily experience; it is the *principle of life*, or *vital affinity*, which binds together the chemical elements in certain forms or *groups*, which are nowhere known but in organized, living beings. Between these two kinds of affinities, then, — the natural chemical affinity on the one hand, and the *vital affinity* on the other, — there is a constant struggle, the one to counteract the operations of the other.

Perhaps this question may arise in your minds, as it has with me, namely, Why should inorganic chemical affinities be called the natural affinities, any more than those which are exhibited by organic life? Is not one as extensive in its influence as the other, and does not the vital affinity, in assimilating material for organized bodies, tend just as much to decompose bodies held together by the natural chemical affinities, as in the reverse way? I cannot anticipate what may be thought of this question by the physical causists, those who maintain that life, organized bodies, originate through the operations of physical agencies, or in other words, the natural chemical affinities. Can the latter transform themselves into vital affinities? It may be so, if the two differ from each other only in degree.

But to return to our proposition, which is that *all living beings* are composed of four chemical elements, and these elements are held together by the *principle of life*. We cannot go behind the veil to see the ultimate condition of this principle;— it must remain, like the principle of gravity, of electricity, and many other things, a partially unexplained phenomenon;— suffice it for our purpose that we consider it perhaps as fully understood as is ordinary chemical affinity. A drop of water is composed of two elements, Oxygen and Hydrogen, represented by O H, which are held together by ordinary or natural chemical affinity. Common hartshorn or ammonia is composed of two elements, namely, Nitrogen and Hydrogen, in these proportions, $N H^3$. Alcohol is composed of three elements, namely, Carbon, Hydrogen, and Oxygen, represented by $C^4 H^6 O^2$, in the proportions indicated by the figures, and these also are held together by natural chemical affinity. This is easily understood. It is a very simple proposition.

Now, in place of this so-called natural chemical affinity, substitute in your minds that other kind of affinity which I have called the *vital affinity*, and apply it to these selfsame chemical elements, Carbon, Hydrogen, Oxygen, and Nitrogen, C H O N, and you have an organized being, a plant or an animal. We may have on one side of a line, *life*, C H O N, and on the other side the same elements, C H O N, but in different proportions, representing the *absence of life*, which is death, and between them, *circumstances*[*] which determine the conditions of these elements, whether they shall exist in one combination or another.

*Organic life* = C H O N, circumstances, *Inorganic bodies* = C H O N.

Without reflection this might seem like the exchange of one chemical compound for another, and not the substitution of life for death, or the substitution of the vital affinity in place of natural chemical affinity; but nevertheless it is true that so

---

[*] Thus, for instance, flesh or blood is composed of $C^{48} H^{39} N^6 O^{15}$, and when these decay and putrefy, Carbonic Acid ($C O^2$), Water (H O), and Ammonia ($N H^3$), are the result.

# ORGANIC LIFE. 9

closely do these two affinities approach each other *in the range of their actions*, that there is on the one hand the drop of resin, gum, or mucus, held together by the natural chemical affinity, and on the other hand, there are certain *living beings* so exceedingly simple in structure that they may be compared to a drop of gum or mucus, but from which they are distinguished by being held together and *animated* by the affinity which is called the *principle of life*. These creatures are so simple that under the most powerful microscopes of the present day they appear like drops of gum or starch, and in fact they would not be recognized as living beings, — I mean to say, in the most literal sense, that they could not be distinguished from the gum-drop — did they not move, and take in food, even living animals as a prey, which they digest in the most simple manner.

The creatures to which I refer are commonly known as the
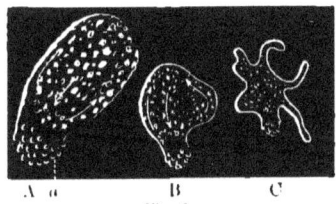
Fig. 1.
Protean animalcules, on account of their changeable form. This diagram (fig. 1) represents the one which is called by scientific men, *Amœba*. The three figures represent the various forms which I have seen the same individual assume, whilst I had it under the microscope, as it crept over the water-plants upon which it is accustomed to dwell. The most usual form which it assumed is that of an elongated oval (A), but from time to time the sides of its body would project either in the form of simple bulgings (B), or suddenly it would spread out from several parts of the body (C), as if it were falling apart; just as you must have seen a drop of water do on a dusty floor, or a drop of oil on the surface of water; and then again it retracted these transparent arms and became perfectly smooth and rounded, resembling a drop of slimy, mucous matter, such as is oftentimes seen about the stems of aquatic plants. But what gave character to this

Fig. 1. *Amœba diffluens*. Ehr. The Protean animalcule. 100 diam. A, B, C, the three attitudes of the same individual. *a*, the head. The arrows indicate the course of the circulating food.—*Original*.

unstable globule, in all its changes of form, was that it invariably progressed with a certain part of its body forward; and upon close examination I found that this part had a peculiar appearance which distinguished it from the rest of the body, that is, it was covered with little knobs (*a*), and the interior was perfectly clear and transparent, with not the least trace of motion to be discovered therein. This I shall call the head. This was in marked contrast with what was to be seen in the other part of the body, for there were the most unmistakable signs of life, exhibited not only by the activity of the numerous particles of food, but by the regularity with which they circulated in currents, having a fixed direction. The arrows which I have introduced in the figures (A B) will indicate the trend of these streams. Constantly and invariably the currents of food passed backwards along the sides of the body to the posterior end, and there uniting in a single stream, turned their course toward the head, passing along the middle line of the body, and again turned off right and left backwards along the sides. Nothing could be more wonderful than this sight; not a sign nor trace of any interior organization which could be supposed to direct these currents; all was as clear as crystal, excepting the circulating particles; and they seemed as if impelled by magic. To increase this almost unavoidable illusion, there was another phenomenon which was connected with the movements of the circulation, which I could not explain to myself for some time; I mean, that, with the exception of the head, the whole of the body seemed to be gradually sliding on itself, and turning over, end for end. This I knew could not be strictly true; and yet the deception to the optical senses was perfect, even after I had discovered the cause of it. This is akin to what every one must have observed whilst sitting in the cars at a station, when a passing train gives to the one you are in the appearance of moving; and this cannot be banished from the impression until the eye is cast upon some fixed object. The illusion in the case of the Amœba was derived from the fact that the particles of food, as they passed backward along the sides of the body, pressed so closely to its surface as to project it more or less, and thus they appeared to be a part of the body itself. The food, which may be either a living animal or plant, is evidently introduced into the body at any point, simply by being adhered to by the viscous surface, and gradually engulfed in the transparent *sarcode*, as this sort of structureless animal-tissue has been called.

Now what I wish particularly to draw your attention to, in this somewhat minute description of the Amœba, is that so exceedingly simple a structure should perform such a variety of acts. It creeps and changes its form, which indicate a muscular power; and seeing that one end of the body always precedes the other, it is fair to draw the inference that this muscular power is under the directing control of at least a certain degree of nervous sense. And again the introduction, circulation, and digestion of food, and the final rejection of the harder indigestible parts of the prey, all point unquestionably to a function which is proper to animals and not to plants. There can be no doubt, then, that this particle of slime-like matter, which is called Amœba, is an animal in the fullest sense of the term.

These other three diagrams (figs. 2, 3, 4,) are intended to illustrate the manner in which the complication of the organization is brought about, as we rise from the lowest forms of life, as represented by the Amœba, through the gradually more elevated types to the highest in this peculiar group of beings. The first step in this advance is made by the addition of a covering, of such a form in this case as to restrict the prolongation of the protrusile parts of the body to one region. In the figure before us, which is that of a Difflugia, (fig. 2,) the body is enveloped in a globular or pear-shaped membranous sac (*a*)

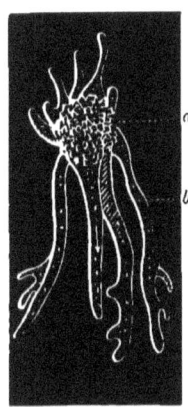
Fig. 2.

Fig. 2. *Difflugia proteiformis*. Ehr. 100 diam. A view from above, looking down upon the top of the shell, *a*. *b*, the pseudopodia, projecting from before and behind. The arrow indicates the flow of the granulated fluids of the body. — *Original*.

to which angular grains of sand are adherent. This has been secreted, or rather excreted by the surface portion of the animal, and the minute fragments of sand are stuck on by some unknown process. At one side of the sac there is an aperture, through which alone the body can protrude. This is not to be seen in the figure, because it is on the opposite side of the shell to that which is here represented. From this aperture the body sends out processes (*b*) similar to those of Amœba; but they appear to have a more definite character, and a seemingly more especial office than in that animal. These processes have been called *pseudopodia*, i. e. false feet, by which name I shall designate them hereafter. Owing to their exceeding transparency, the pseudopodia, as they stretch out over the surface upon which the animal creeps, remind one of water spreading in streaks over glass, and sending off little branches, here and there, sideways. Sometimes only a single pseudopod is stretched out, and waves gently from side to side as if feeling for something.

One cannot help but admire the caution with which this seems to be done, for after reconnoitering awhile the other pseudopodia come forth. There would seem to be good reason to believe that the animal really does exercise at least an instinctive caution, from the timidity which becomes apparent when it is disturbed; for instantly, and as if with a sudden jump, the body darts toward the ends of the pseudopodia, and the latter being contracted in the act, are then drawn into the shell. The leap thus made is owing to the fact that the contraction of the pseudopodia is more rapid than the loosening of their hold, and consequently they drag the whole body toward the point of attachment.

I mention this phenomenon particularly that I may draw your attention to the rapidity of the muscular contraction which is exhibited in the act. We are apt to suppose that a low degree of organization has a correspondingly low vitality; and this is true to a certain extent; but we are not justified in supposing it to be in exact parallelism with the organic grade of the animal, for here is one instance among many others in which the muscular action of a scarcely organized body is as rapid as in the most highly perfected types of animals.

The circulation of the fluid and granular contents of the body is more active than in Amœba, but it does not appear to be at all different in kind or of a more complicated nature, although it seemingly is so because it has the appearance of being confined in more restricted channels, as it passes along the slender pseudopodia; but these channels are mere hollows with as indefinite boundaries as in Amœba.

From this we may infer that it is not by a general advance of the whole organization that the upward steps, in the development of types, are made; but here and there one organ after another is either added, or more and more specialized in its functions, until, by insensible grades, the highest type of organization within each group is attained.

In this other figure, (fig. 3,) which represents an animal which I kept by hundreds in my marine aquarium for eighteen months, the advance in grade is made by a simple complication of the covering of the body; merely by elongating and coiling the dormitory of the little creature, so as to resemble the spiral shell of some of the snails. The pseudopodia (*b*) are exceedingly transparent, pointed, and so excessively slender toward their tips that it requires the best powers of the microscope to see them; but yet, upon the least disturbance, these frail threads retract their length down to almost nothing, with a lightning-like rapidity; even while you are looking at them there is a sudden shock, and they are gone!

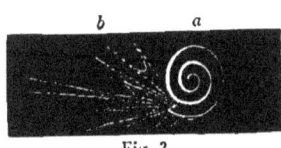
Fig. 3.

How slight indeed is the degree of organization required in which to manifest some of the most active powers of vitality! Almost within a step we have the dazzling complication of the

Fig. 3. *Cornuspira planorbis.* Schultze. 50 diam. Represented as it crept over the glass side of the aquarium. *a*, the shell; *b*, the pseudopodia, partially extended backward over the surface of the shell. — *Original.*

inorganic crystal, revelling in glittering angles of such definite proportions and relations that an infinite design shines forth at every turn; and yet the simple faculty of self-determinative *motion* stamps upon the almost shapeless mass of the Amœba a character by which the mind, as it were instinctively, places it at a far more elevated status than the attractive mineral.

There is but one more figure which I shall introduce to your notice, as I think that the animal which it is intended to illustrate represents within itself the highest tendencies of organi-

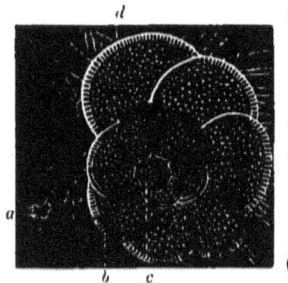
Fig. 4.

zation that may be found in the group of Rhizopods. This animal, (fig. 4,) which was called by Schultze, its discoverer, *Rotalia Veneta*, is to be found crawling over the slimy mud among the lagoons of Venice. From point to point along the turns of the spiral shell (*b*, *c*, *d*) there are transverse partitions, which divide its cavity into as many chambers, but do not shut them off from one another entirely, as there is left a passage-way from one to the other, through which the soft parts of the animal connect with each successive one, from the central globular chamber (*c*) to the broadest and last formed one (*d*) at the edge of the shell. The pseudopodia (*a*) are not restricted in their egress to the single aperture at the termination of the shell, as in the last animal, but they project from all parts of the body through fine pores in the shell, as you see in the figure. It is not necessary, however, that you should infer that the food, whether animal or vegetable, is of necessity very minute in order to be introduced within the shell; for I must tell you that the digestion may go on outside as well as within, and that it is done in quite a simple manner. The pseudopodia, as I have told you, are mere prolongations of the body, and the circulation of nutrient particles extends to their

Fig. 4. *Rotalia Veneta.* Schultze. 72 diam. *a*, the pseudopodia, projecting in every direction through the pores of the shell; *b*, the transverse partitions; *c*, the original, primary chamber; *d*, the last chamber. — *From Schultze.*

very tips. When, therefore, any living thing is seized upon, it is at once enveloped in a glairy mass (a), which is formed by the pseudopodia fusing their sides together; and in this temporary stomach the nourishment is extracted from the victim, and carried in the circulation to the main part of the body.

I think this will suffice to show you what is the extent of the duties which these simple creatures perform. It is true that their functions are not very complicated, but yet they are far more so than any one, knowing their simplicity, would suspect them to be capable of.

Were you to imagine these chemical elements, Carbon, Hydrogen, Oxygen, and Nitrogen, (C H O N,) to be united in the most simple manner, in order to form some animate creature, you could hardly produce a more lowly organized being than these self-same Amœbas which I have illustrated here. They are, in truth, among the lowest of all animals known.

While we are engaged upon this part of our subject, it would seem to be most fitting to introduce here a description of some experiments which were made to ascertain under what conditions these rudimentary forms of life may originate. I will refer to only one set of experiments, because they seem to be, by far, the most satisfactory of any that have been made.

In July, 1862, Professor Jeffries Wyman, of Harvard University, Cambridge, Mass., published, in the "American Journal of Science," a paper whose title runs thus, —

"*Experiments on the formation of Infusoria in boiled solutions of organic matter, enclosed in hermetically sealed vessels, and supplied with pure air.*"

I propose now to make some extracts from that paper, and illustrate what is therein stated by these diagrams. After some preliminary remarks, Professor Wyman proceeds thus: —

" In order that the reader may understand what precautions were taken, we shall first describe the manner in which the experiments were performed."

" (1.) In some instances (as in Expts. i. to v., vii. to xi., xiii. to xv., xxix. and xxx. inclusive) they were prepared as in fig. A.

The materials of the infusion were put into a flask, and a cork $a$, through which was passed a glass tube, drawn to a neck at $b$, was pushed deeply into the mouth of it. The space above the cork was filled with an adhesive cement $d$, composed of resin, wax, and varnish. The glass tube was bent at a right angle, and

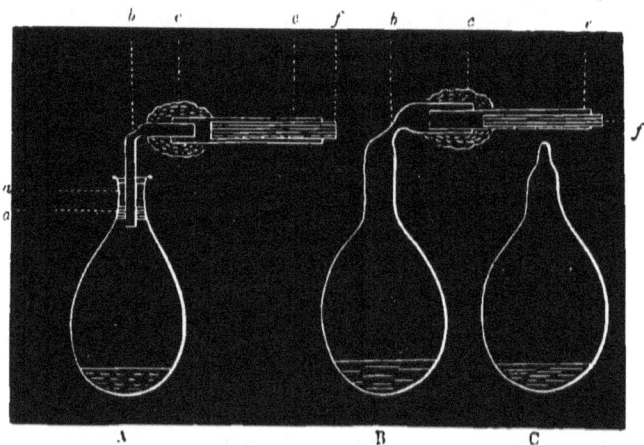

inserted into an iron tube $e$, and cemented there with plaster of Paris $c$. The iron tube was filled with wires $f$, leaving only very narrow passage-ways between them.

"(2) Others (as in Expts. vi., xii., xvi. to xxiii., and xxxi. to xxxiii. inclusive) were prepared as in fig. B, in which the joining at $a$, fig. A, is avoided, and the iron tube is cemented directly into the mouth of the flask, the neck of which is drawn out at $b$, to render the sealing of it easy; otherwise the conditions are the same as in fig. A.

"(3.) In other experiments (as in Expts. xxiv. to xxviii., and xxxiv. to xxxvii. inclusive) the flask, fig. C, was sealed at the ordinary temperature of the room, and submerged during the period of the experiment in boiling water. This was the method followed by Needham and Spallanzani, and has the merit of eliminating all suspicions of error which might be supposed to arise from some imperfections in the joinings.

" In the first and second methods, the solution in the flask is boiled, and at the same time the iron tube filled with wires is heated to redness. While the contents are boiling the steam formed expels the air from the flask; when the boiling has continued long enough, the heat is withdrawn from beneath the flask, and, as the steam condenses, the air again enters through the iron tube, the red heat of which is kept up, so that all organisms contained in the air are burned. In both methods the flask is allowed to cool *very slowly* in order that the entering air may be as long as possible in passing through the iron tubes, and thus the *destruction of its organic matters* insured. When cold, the flasks are *sealed* at *b*, figs. A and B, with the blowpipe.

" In experiments xxix. and xxx., a glass tube filled with asbestos and platinum sponge was used instead of the iron tube filled with wires.

" The time during which the infusions were boiled varied, as will be seen by the records, from fifteen minutes to *two hours*, and the amount of infusion used was from one twentieth to one thirtieth of the whole capacity of the flask, the object being to have the materials exposed to as *large a quantity* of air as possible.

" In the account which follows, especial mention is made, in most instances, of the time of the formation of the '*film*.' This is always the first indication which can be had, without opening the flasks, that minute organisms are developed; it is in fact made up entirely of them, as has been proved by repeated examinations with the microscope.

"After the flasks were prepared they were suspended from the walls of a sitting-room, near the ceiling, where they were exposed to a temperature of between 70° and 80° F. throughout the day and nearly the same during the night.

" Expt. xii. (B.) * March 13th. The juice of an ounce of beef, to which was added 10 cub. cent.† of urine and 40 c. c. [cubic

---

\* " The figure [letter] in brackets following the number of the experiment indicates which of the three modes of preparing the experiment was made use of."

† Cubic centimetres. A centimetre is about equal to $\frac{2}{5}$ of an inch, and therefore a cubic centimetre is equal to the cube of $\frac{2}{5}$ of an inch, which is $\frac{8}{125}$ of a cubic inch.

centimetres] of water, was boiled 20′ [20 minutes] in a bolt-head and hermetically sealed. A film formed on the fourth, and the flask was opened on the eleventh day, when there was a distinct rush of air outwards. Large numbers of Bacteriums (fig. 6, *a, b, c, d*) were found, also small spherical bodies, (fig. 5, *a*,) with ciliary motions and oval bodies like Kolpoda, containing what appeared to be Bacteriums; one of these Kolpoda-like bodies moved with cilia. (Fig. 5, *c, d*.)

"Expts. xvi., xvii., xviii., xix., (B,) March 20th, were made with juice of beef and water in flasks of 550 cub. cent. capacity;

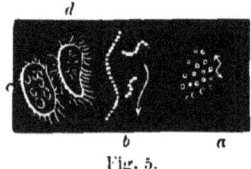

Fig. 5.

Fig. 5. *a*, a group of *Monads;* minute spherical bodies, constantly agitating, as if moved by the vibrations of thread-like appendages (cilia); *b*, *Vibrio rugula*. 500 diam. A very common thing in all decaying fluid matter, forming a sort of scum at a certain period of the decomposition. Seen with the lower powers of the microscope, the scum appears to scintillate all over its surface; and if higher powers are applied, this phenomenon is found to be due to the rapid whirling motion of these minute, curved, moniliform (bead-like) strings, as they shoot, with greater or less velocity, backward and forwards across the vision. Their curved form produces the appearance of a screw or spiral as they vibrate in their path through the fluid, and the alternate appearance and disappearance of the highly refractive bead-like grains has the effect of sudden flashes of light, or scintillations. When there are only four or five beads in a string, the velocity of their movements is almost inconceivable, and their structure cannot be ascertained until they stop to change or reverse the whirl, and return upon the path in which they came. Most frequently, however, there are six or eight beads in a string; and in this condition, although their speed is less than that of the shorter ones, yet they swim very rapidly; but as the string is found to be longer, so do we see also that the motion is proportionately slower; and finally we may meet with those very long ones which wind their way with the leisurely undulations of a snake.

*c, d,* "Kolpoda-like bodies," seen in Expt. xii. The presence of movable cilia on a body is the most indubitable evidence that it is in a *living* condition; but whether it is an animal or plant must be determined by its internal structure. Professor Wyman's unpublished figures, from which these two were copied, by his kind permission, do not indicate an animal any more than a vegetable nature. They would seem to be allied to those extremely low forms of life, whose position either in the animal or vegetable kingdom is to be hereafter determined by more searching investigations than have thus far been made.

xvi. was boiled 15', the film formed on the second day and the flask was opened on the ninth. Vibrios were found in abundance, (fig. 5, b,) of different lengths, some of them moving with great rapidity. xvii. was boiled 30', the film was formed on the third, and the flask was opened on the ninth day. Vibrios were found in great numbers, some of them bending and extending themselves rapidly. Some minute spherical bodies were also seen, having the kind of motion which results from vibrating cilia, though none of these were detected. xviii. was boiled 15', the fluid having been previously filtered; the film formed on the third, and the flask was opened on the eighth day: the organisms found were the same as in xvii. xix. was boiled one hour. The film formed on the second, and the flask was opened on the twenty-fourth day. The infusion had a slightly putrid odor, and contained Vibrios and Bacteriums.

" Expts. xxix., xxx. (A.) February 17th. In both of these the contents of the flasks were solutions of sugar and gelatine in water, to which fragments of cabbage-leaves were added. The air was introduced through a *Bohemian glass tube*, filled with *asbestos and platinum sponge*, and heated to redness. The materials were boiled 30'. In xxix. the film was formed on the

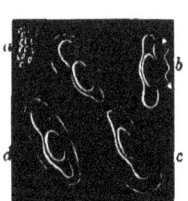

Fig. 6. *Bacterium (Zoöglœa) termo.* a, a group magnified 500 diameters; b, c, d, magnified 3500 diam. When seen as at a, they appear like minute, oval, brilliantly refracting granules, dancing in a constant zigzag, something like a swarm of mosquitoes vibrating in a sunbeam. They never move in a direct course, but ever hover about the same spot; in this respect imitating the spores or seeds of certain aquatic plants. By the application of a higher power of the microscope we may recognize their shape more distinctly. Most frequently they were found, as at b, simple guitar-shaped bodies with a dark oval centre, and excessively transparent ends; but now and then one or two of them were enclosed in an exceedingly transparent envelope, as at c and d, which appeared like a halo around them. In this condition they resembled some of the jointed Vibrios. The transparent gelatinous envelope is eminently characteristic of certain kinds of mould, growing in damp places. — *Original.* (See fig. 8.)

Fig. 6.

twenty-ninth, and the flask was opened on the thirty-ninth day. The solution was found to contain Bacteriums, and cells filled with them. In xxx. the film was formed on the seventh day, and Bacteriums were found on the twenty-third, when there was a slight odor of putrefaction.

"Expts. xxxi., xxxii., xxxiii. (B.) March 24th. 30 grains of sugar, 20 c. c. of beef-juice, 158 c. c. of water, were divided into three parts, and each part put into a flask of 550 c. c. capacity, and boiled 15'. No film was formed in either of them. xxxiii. was opened on the thirtieth day; *ferment cells* (fig. 7) and some

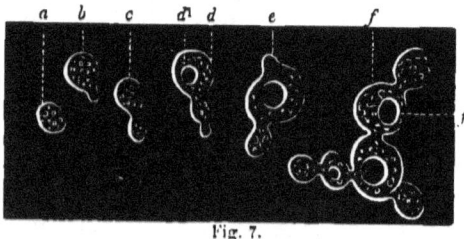

Fig. 7. *Torula Cerevisiæ. Yeast-plant.* Mag. 500 diam. *Original.* From a new yeast. In all fluid fermentations, in yeast, and among decomposing fluid matter, what are called "ferment-cells" are found in greater or less abundance. It is in such conditions that this lowly organized member of the vegetable kingdom finds its proper basis for origin and development. Where it is abundant one may have all stages of growth represented at one time in the field of the microscope, as they are illustrated from $a$ to $f$, in this cut; and by watching for a few hours the whole process of cell multiplication from $a$ to $f$ may be seen. The little spherical granulated cell at $a$ is about $\frac{1}{4000}$ of an inch in diameter; but it was originally much smaller, for it may be traced from excessively transparent globular bodies, with soft, delicate outlines, not more than $\frac{1}{10000}$ of an inch in diameter. As it progresses in development, one side of the cell bulges out, as at $b$; and this bulging grows until another cell is formed like the one at $c$; then the first cell, increasing in size, ($d$.) develops another cell-like body, ($d^1$,) called the *nucleus*, in its fluid contents, whilst the second cell sends out a bulging process, which eventually becomes a third cell. In this way a single string of cells is formed; but quite as frequently the primary cell develops a new cell from two different points, as at $e$; or even three cells are developed from the primary one, (see the largest cell at $f$,) and, each of these secondary cells developing a cell from its sides, produce together the irregular branching form, which we have here represented as the perfect "yeast-plant" ($f$). It is well known now, however, that this is not its perfect state. Under favorable circumstances, the yeast-plant, so-called, rises to the surface of the fermenting fluid, and clinging to the side of the vessel, allows its growing

filaments of a doubtful vegetable appearance were found. xxxii. was opened on the forty-second day, and contained *ferment cells* and *monads* (fig. 5, *a*). An escape of gas took place when the flask was opened. xxxi. was opened on the forty-third day, and found to contain *ferment cells* in large numbers, *in different stages of cell multiplication;* as in xxxii., there was an escape of gas.

" Expt. xxxiv. (C.) March 27th. Juice of mutton, in a hermetically sealed flask, was boiled 5' in a Papin's digester, under a pressure of two atmospheres.* A film formed on the fourth day. It was opened several days later, in the presence of Professor Gray, and found to contain *Vibrios*, (fig. 5, *b*,) and *Bacteriums*, (fig. 6,) *some of them moving with great rapidity.*

" Expt. xxxv. (C.) The same as the preceding, and boiled in Papin's digester 10' and under the pressure of five atmospheres.† No film was formed. The flask was opened on the forty-first day. Monads, (fig. 5, *a*,) and *Vibrios*, (fig. 5, *b*,) *were found, some of the latter moving across the field.* No putrefaction ; the solution had an alkaline taste.

" Expt. xxxvi. (C.) March 28th. Beef-juice was filtered and boiled, as in the preceding experiment, 15', under two atmospheres. Opened on the forty-first day, and no evidence of life was found. When the end of the flask was heated, previously to opening, it collapsed.

" Expt. xxxvii. (C.) March 28th. The same as the preceding; boiled 15' under five atmospheres. Opened on the forty-first day, and no evidence of life was detected.

branches to project into the air. The emergence from its fluid habitat is the beginning of another stage in its growth ; and in fact it could not otherwise perfect its development. In course of time, then, its branches become a tangled mass of white threads and bristling points. This is the condition in which it is known as " white mould"; or, when the bluish or greenish spores (seeds) are ripened, on the ends of the bristling points, " blue mould."

* Two atmospheres are equal to $250.52°$ Fahrenheit, or about $38°$ above boiling point.

The pressure of five atmospheres is equivalent to $307.5°$ Fahr., or about $95°$ above the boiling point.

"We have here a series of thirty-three experiments, prepared in different ways, in which solutions of organic matter, some of them previously filtered, have been boiled at the *ordinary pressure* of the atmosphere for a length of time, varying from 15 minutes to 2 hours, and exposed to air purified by heat.

"In many instances, a solution like that in the sealed flasks, and boiled for the same length of time, was exposed to the ordinary air of the room, in an open flask. Although the same forms were found in the two, they appeared much more rapidly in the open than in closed vessels, and the contents of the former soon became putrid, while those of the others, at the time of opening, were mostly not, and in a few instances only slightly so.

"We have, in addition, four experiments: namely, xxxiv., xxxv., xxxvi., xxxvii., made under *increased pressure*, and sealed by the third method; xxxiv. and xxxvi. were boiled 5′ and 15′ respectively, under two atmospheres, and xxxv. and xxxvii., under five atmospheres for 10′ and 15′ respectively. Evidence of life, consisting of Monads and Vibrios, was found in xxxiv. and xxxv., but none in the others.

"The result of the experiments here described is, that *the boiled solutions of organic matter made use of, exposed only to air which has passed through tubes heated to redness, or enclosed with air in hermetically sealed vessels and exposed to boiling water, became the seat of infusorial life.*

"The experiments which have been described throw but little light on the *immediate* source from which the organisms in question have been derived. Those who reject the doctrine of spontaneous generation in any of the forms in which it has been brought forward, will ascribe them to spores contained either in the air enclosed in the flask, or in the materials of the solution.

"Those who advocate the theory of spontaneous generation, on the other hand, will doubtless find, in the experiments here recorded, evidence in support of their views. While they *admit* that spores and minute eggs are disseminated through the air, they assert that *no spores or eggs of any kind have been actually proved by experiment to resist the prolonged action of boiling water.* As regards Vibrios, Bacteriums, Spirillums, etc., it has not yet been shown that they have spores; the existence of them is simply inferred from analogy. It is certain that Vibrios are killed by being immersed in water, the temperature of which does not exceed 200° F. We have found all motion, except the Brownian, to cease even at 180° F. We have also proved by several experiments that the spores of common mould are killed, both by being exposed to steam and by passing through the heated tube used in the experiments described in this article. If, on the one hand, it is urged that all organisms, in so far as the early history of them is known, are derived from ova, and therefore from analogy we must ascribe a similar origin to these minute beings whose early history we do not know, it may be urged with equal force, on the other hand, *that all ova* and *spores*, in so far as we know anything about them, *are destroyed by prolonged boiling:* therefore from analogy we are equally bound to infer that Vibrios, Bacteriums, &c., could not have been derived from ova, since these would all have been destroyed by the conditions to which they have been subjected. The argument from analogy is as strong in the one case as in the other."

On the 4th of August, 1863, Prof. Wyman put into my hands, for examination, the contents of a sealed flask which he had just opened, and which was prepared according to the method (B) on the 22d of July previous, *i. e.* thirteen days before. I directed my attention particularly to the structure of the Bacteriums, (fig. 8, *a*, *b*,) which floated in immense numbers throughout the fluid. Ordinarily, the central part of the guitar-shaped body is occupied by a dark oval mass, as I have described it in fig. 6, but in this examination, by the help of an immensely

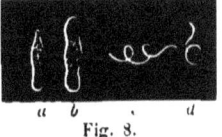

Fig. 8.

Fig. 8. *a*, *b*, *Bacterium (Zooglœa) termo*. Dnj. 3500 diameters. Two of the most diverse forms, with a granulated centre and transparent ends. — *Original.* *c*, *d*, *Spirillum.* Spiral thread-like bodies oftentimes seen among decaying substances in fluids. They whirl with great velocity. — *Original.*

magnifying lens,* that gave me a power equal to 3500 diameters, I found that this centre was composed of a number of granular bodies, (fig. 8, *a*, *b*,) but the ends of the Bacterium were, like those in fig. 6, very transparent; and, moreover, what gave additional character to these bodies, one end was always much more delicate than the other, at times so excessively faint as to be scarcely discernible. What induced me, among other things, to believe that these bodies were in a peculiar stage of growth, was, as Prof. Wyman had noticed at the same time, that they did not move, as they are ordinarily observed to do.

From the foregoing observations it is clear that the Bacteriums, found in the sealed flasks, are of a more elevated nature than the simple granules which go by the name of Monads; in fact, they are even more highly organized than the Vibrio, which is essentially nothing more than a string of Monads moving in concert. Finally, I will add, in confirmation of what I have said in regard to Bacterium, that a German observer, Cohn,† has shown that the *Bacterium termo* is merely one of the stages of growth of a kind of mould which he has called *Zoöglæa termo*.

In one of Prof. Wyman's flasks, containing some slivers of beef, sugar, and water, which he had prepared, according to the method marked (B), and boiled 20′ on the 2d of Sept., 1864, and which I opened, at his request, on the 25th of Oct., 1864, I found large numbers of *Bacterium termo* oscillating very lively. These I have illustrated and described under fig. 6. The moniliform Vibrio (fig. 5, *b*,) was seen only here and there; but there abounded another form of Vibrio, which, when seen with a magnifying power of only 300 diameters, might be mistaken for the moniliform Vibrio; if, however, they were magnified with a good lens, 500 diameters, their true outlines would become apparent. In order to get as clear a conception of their form and nature as possible, I subjected them to the searching scrutiny of

---

* One of Tolles's $\frac{1}{25}$ of an inch objectives, made for me with particular reference to the study of the minuter Infusoria and the early stages of cellular development.

† See Cohn, Acta Academiæ Naturæ Curiosorum, 1854; vol. XXIV. pars I. p. 119, and Pl. XV. fig. 9.

a $\frac{1}{25}$ of an inch lens. With this power, of 3500 diameters, I made out distinctly that the *Vibrio baccillus*, (fig. 9,) as it is called, consists of a series of little rods which are joined end to end by a delicate membrane; or rather, I might say, that they are ensheathed in a tubular membrane, with sufficient space between the successive rods to allow them to double upon each other. Although they are represented here under a lower amplification, — only 2000 diameters, — yet I think it is sufficient to show the manner in which they are connected with each other. Their movements are very much like those of *Vibrio rugula*, but still they have a method of progression which is eminently characteristic, for such minute bodies, when seen with the highest magnifying powers. I know of nothing so apt to compare them with, when in motion, as a jointed toy-snake vibrating. The effect is most striking when one of the longest, many-jointed specimens moves across the field of the microscope with a sort of disjointed action, as if each rod held an independent course in one common stream. Sometimes a considerable portion of the end of the sheath was empty and collapsed, (fig. 9, c,) and in this condition, being quite flexible, it was difficult to persuade one's self that it was not a vibrating cilium, as it waved from side to side during the undulating progress of the chain. This, as well as the moniliform Vibrio, is generally considered to be allied to certain aquatic, filamentous plants, common in our streams and ponds, which are known as Oscillatoria.

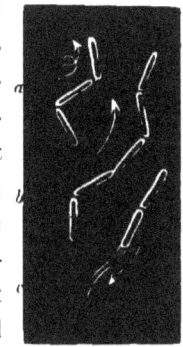

Fig. 9.

Now it is true that in these experiments of Professor Wyman, the matter which was introduced into the flasks was not reduced to its separate chemical elements C, H, O, N, but rather to the fluid state. This is the condition of animal and vegetable substances when in a decomposed state; and when such large numbers of animalculæ, identical with those which have just

Fig. 9. *Vibrio baccillus.* Ehr. 2000 diameters. *a, b, c,* three individuals in various attitudes of flexion whilst in motion. — *Original.*

been described, develop in the putrid mass. If left to the action of time the whole fluid would eventually resolve itself into the ultimate chemical elements C, H, O, N, (see pages 7 and 8); but this was not desirable in these experiments. Under these decomposing conditions we may see one of Nature's modes of reproducing or rather generating new forms of life by *spontaneity*.

Professor Wyman has since been so kind as to show me some other experiments which prove that these same bodies which are developed in the sealed flasks, are killed at a point far below that of boiling water. He placed large quantities of each kind, viz: *Vibrios, Bacteriums, Kolpodas*, &c., each in a separate test tube, and each tube with a thermometer in a water-bath, and applying heat to the bath, he examined, from time to time, as the temperature was raised, portions of each set with the microscope. Some of the animalculæ survived the heat up to 125°, and others up to 130°, but in no instance did any of them live when the temperature was raised to 150°, which is sixty-two degrees below boiling, 212.° You will recollect that in all of the experiments with the sealed flasks, the fluid was raised to the boiling point, 212°, and in some of them it was raised still higher, in one case to 250°, which is 38° above boiling, or 100° above the temperature at which these animalcules can live. In another instance the flask was heated to 307°, which is 95° above boiling, or 157° above what these creatures can live in.

The fact that the experiments with the *sealed flasks* proved, — if anything can be proved beyond the reach of change or improvement, — that beings with motion, undoubted living beings, were produced where life could not possibly have existed previously, is a sufficient basis for the further assumption that still higher forms could arise from these. That is to say, if, under the conditions arranged in the sealed flasks, living beings, either animals or plants, of the lowest degree arise, there is nothing illogical in assuming that from these lowly organized, *animate* bodies somewhat higher and more complicated beings may originate. Keeping these experiments in view let us now return to the Amœbas and their protean congeners, Difflugia, Cornuspira, &c.

Calling to mind now what I have said about the extreme simplicity of the lowest animate forms, the Amœbas, Difflugias, and the like, let us turn to an examination of the *lowest condition of life in which any animal, whether high or low, can exist.* I mean *the egg-state.*

## CHAPTER II.

WHAT SPONTANEOUS GENERATION PROVES. — THE EGG IS THE LOWEST PHASE OF ANIMAL LIFE. — THE EGG A BIPOLAR ANIMAL. — THE EGG CONTRASTED WITH THE LOWEST FORMS OF ADULT BEINGS. — THE EGG AS RELATED TO SECONDARY CAUSES.

AT the end of my last lecture I announced that I would next take up the consideration of the *egg* of animals, as that is the phase in which *all animals*, at one time, are in the *lowest possible condition of life*. Before I proceed to do this, however, I wish to preface it with some remarks upon the theory of *spontaneous generation*, which may serve not only as a recapitulation of what I have already said upon that subject, but will, I hope, make more clear to your minds the object for which the experiments in the sealed flasks were instituted; and it will at the same time lead in a direct line to the consideration of what may be the relation of the *egg-condition*, of all animals, to the adult state of the lowest forms of animate beings.

When geologists first announced that the earth contained the remains of animals and plants, which, from the nature of the action of physical causes, and the position in which these fossils were found, demonstrated that they belonged to a period, or series of periods, anterior to the creation of man, the assertion was received, by what was then called the "Church Party" in England, with an expression of horror, that scientific men should attempt to support atheism by the palpable evidence of scientific facts! Without stopping at the present time to show how this holy horror was finally, and in the process of time, changed to an enthusiastic and religious advocacy of this geological theory, I will merely state the fact, and then pass on to the result of this general recognition and admission of the existence of a pre-Adamite world.

Having made this admission, it became evident at the same time that a series of creations had been going on previously to the origination of the present race of beings on the globe; and out of the contemplation of this idea arose the question as to whether the Creator has not *continued* to exercise the creative faculty at *all times*, even to the *present day*. They recognize the Creator's controlling hand when they see the child resembling the parent; it is not blind chance to them that like comes from like; there must be some reason for this, and the reason they give is that the Creator is continually active in the administration of his laws. If, therefore, he is visibly present in the operation of one series of acts, it may be that he *still continues* to carry on another series, which we know he has at some time in the remote past *begun*. At some unknown, distant period, animals originated through creative influence on this globe.

Now, then, the question arose in this form, namely, did all animals, which have appeared from the beginning, originate by *birth* from the first created; or is this only *one mode* of continuing their presence on the earth, and has the Creator also *constantly repeated* his original creation through all time, even until *now*? This question arose from various reasons, and among others, because observers had noticed the fact, that, when a pond or stream, or a whole tract of country, dries up, as oftentimes happens in the summer months, as a natural consequence, all the animals and plants in it, which are dependent upon water for their existence, die for want of their natural element; but when the rains of autumn have refilled these streams and ponds, the aquatic animals appear again. This some observers accounted for by supposing that the eggs and seeds of these animals and plants were constantly floating in the air, and were washed down by the rains, and thus the ponds became restocked with life; but other observers disputed the fact that there were seeds and eggs floating about in the air, or insisted that they were in such small numbers that they could not possibly account for the sudden appearance of such large quantities of living creatures in these ponds; and therefore they propounded the idea that they originated there exactly in the same way as did the first aquatic animals that originally began life in the rivers and ponds of this globe; that is, they were created there.

In support of this theory they sought to contrive some way by which they might construct artificial ponds which should be shut off from the surrounding air, and therefore no floating or flying seeds or eggs, if they really existed, could gain access to these isolated waters. There have been several contrivances set up which were destined to carry out this plan, and among others, the one which I have already described to you. The various kinds of fluids, such as beef-juice, mutton-juice, sugar-water, ammonia, gelatine, &c., which were introduced in the different experiments, were for the purpose of furnishing a diversity of conditions and food for the creatures which might originate therein, and also to imitate the decomposed contents of stagnant pools and ditches; the heat which was applied to the flasks was to kill all life that might be in the fluid; and the air was allowed to enter the flasks, as they cooled down, through red-hot tubes, in order that whatever seeds or eggs there were floating in the air might be killed by the heat, and the air then would be pure.

In the hermetically sealed flask, C, (p. 16,) the air and the fluid are sealed up before heating, so that after boiling in the Papins digester, this little world within the glass cools down without the least possible chance of communication with anything external to it.

Under these circumstances, the advocates of the continued creation of animals to the present day, claim, that, if living creatures do appear in these isolated pools in the flasks, they must have originated there without the previous intervention of a parental form, because all life had previously been extinguished by the heat. From this, then, they argue, that, when the dried-up pools and streams are refilled by the autumn rains, the animals which appear therein *originate on the spot*, in the same way as did the first animals that peopled this world.

As I said before, at the end of my last lecture, the fact that the experiments with the sealed flasks proved that motile beings, undoubted living beings, originated where life could not by any means have existed previously, is a sufficient basis for a further assumption that still higher forms could arise from these. Keeping in mind now what has been said about the extreme simplicity of the lowest animate forms, the Amœba, Difflugia, and others like them, let us turn to a consideration of the *lowest condition of life in which any animal, whether high or low, can exist. I mean the egg-state!* Let us compare this preliminary, scarcely organized, low form or state of life which we find in the egg, with these gum-drop like beings, as I have called the Amœbas, Difflugias, &c. In the first place we must consider what an egg is, and under what conditions it originates and exists.

Originally, that is when beginning to form, an egg is a very minute aggregation of fluid matter,—more simple even than the Vibrios and Bacteriums of the sealed-flask experiments of Dr. Wyman,—and yet this fluid is gradually transformed into minute granular bodies, the yolk-granules so called, and these in their turn eventually become *cells*, and by combining form a *living, sentient being.* Now when we turn to the lowest animals, such as Amœba, Difflugia, Rotalia, &c., we find among them those

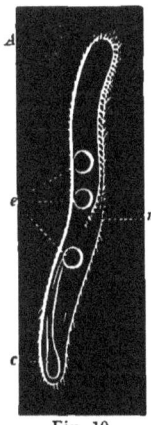

Fig. 10.

which are nearly or fully as simple as the eggs which they themselves lay. In the latter instance, the egg, in order to become an adult, merely changes its form without developing into an appreciably more complicated state. Those undoubted animal infusoria, such as Paramecium (fig. 96), Stentor (fig. 30), Epistylis (fig. 95), Pleuronema (fig. 90), &c., arise from eggs which are not such as might answer fully to the *theoretical egg* of physiologists. Balbiani (Journ. Physiol., 1861, IV.) found that the egg

Fig. 11.

of Spirostomum (fig. 10, A) and Stentor (fig. 11) is a mere cell, without any other sign of the characteristic nucleus-like

Fig. 10. *Spirostomum teres.* Clap. 150 diam. View of the lower side, showing the narrow oblique furrow which passes from the mouth, *m*, to the end of the body ; *c*, tubular contractile vesicle ; *e*, eggs. *A*, one of these eggs more highly magnified. — *From Balbiani.*

Fig. 11. The egg of *Stentor cœruleum.* Ehr. 500 diam. The central clear space is all that represents the germinal vesicle. — *From Balbiani.*

vesicle, the so-called *germinal vesicle*, than a clear spot in the midst of the yolk granules. Were it not for the subsequent development of these, Balbiani could hardly have determined their true nature.

If, therefore, such simple cells are really eggs, we must of necessity form a new diagnosis of the characteristics of the theoretical or *typical egg;* and that I have, for one, been long inclined to do. For the last five or six years I have felt that my studies were leading me to look upon the relations of the various regions of the egg as those of *degrees,* which shade off, the one into the other.

Our ideas of an egg have been heretofore based upon the structure of the eggs of the higher animals, instead of upon what is common to all eggs. For instance, some eggs of the higher animals (fig. 12) have a germinal (or Purkinjean) vesicle; within that a germinal spot (or Wagnerian vesicle), and within the last a nucleolus; all equally and sharply defined, like so many hollow concentric spheres, each and every one having its peculiar, characteristic feature. Under this guise one would hardly suspect the true relations of these parts to each other. But these I will explain presently. In the eggs of some other animals (fig. 13) there are but two of these concentric vesicles, (namely, the Purkinjean *p*, and the Wagnerian *w*,) the innermost one being absent. Again there are those which have but one vesicle: for

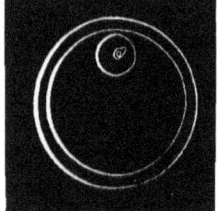

Fig. 12.

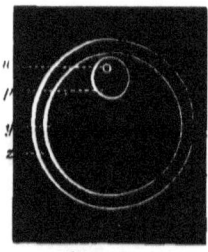

Fig. 13.

Fig. 12. The egg of the *Sow.* 166 diam. Natural size $\frac{1}{200}$ of an inch. The next to the outer circle forms the boundary of the yolk. The small circle at the upper side is the germinal or Purkinjean vesicle; the oval spot is the germinal dot or Wagnerian vesicle; and the central spot is the nucleolus. — *From Thompson.*

Fig. 13. Egg of the *Rabbit.* 166 diam. *z,* the "zona pellucida" or yolk envelope; *y,* the yolk; *p,* the Purkinjean (germinal) vesicle; *w,* the Wagnerian vesicle or nucleolus. — *From Coste.*

instance that of Laomedea, (fig. 14,) a kind of
Hydra. Finally, we come to the lowest degree,
in which, as I have already described, the egg
(figs. 10, 11) is a *mere cell* with a light spot in
one part of the yolk. The eggs of Amœba are
also in the same condition as the last. In this
case the parent of the egg is as simple in structure as the egg
itself.

Fig. 14.

What, then, are the characteristics of an egg? If the sharply
defined, concentric vesicles of the most complicated eggs are
not to be found everywhere, but on the contrary we see all
grades of definiteness and number as regards these apparently
special bodies, we must come to the conclusion that each vesicle
is not restricted within its own boundaries, even though its outline may have the appearance of a wall which would seem to
be intended to shut off direct communication with the outside.

I think that the most straightforward solution of this problem
is, that the *germinal vesicle,* — which is always present in some
form, either sharply defined, or, as I have shown, as a mere
transparent spot at one side of the egg, — is simply an expression
of the *concentration of albuminous matter at one pole;* whilst at
the opposite, or, as one might call it, the *negative pole,* we have
the *mass of yolk.*

That this is so, is demonstrated by the process of development of the egg from its inception to its completion; and as the
egg is a cell, and the type of all cells, its mode of genesis is
typical of all free-cell development. This I will illustrate by a
series of ideal figures (figs. 15, 16, 17, 18) of the progressive
stages of development of the theoretical egg; such a one as
would, in the progress of growth, pass through all the conditions
in which the egg has been known to exist in various animals.
As I have already said, the egg in its inception is a minute
aggregation of fluid matter; but this drop of fluid has not a
homogeneous, uniform density throughout; on the contrary, it

Fig. 14. Egg of *Laomedea amphora.* Ag. 125 diam. The Purkinjean vesicle
is filled with transparent, coarse granules. — *Original.*

makes its first appearance in the egg-bearing organ, the *ovary*, in the form of an indefinitely bounded globule, with a greater degree of transparency on one side than on the other (fig. 15). This diversity in the degree of refraction between the two sides is owing to the difference in the nature of the constituents of the globule; on one side the substance is pure albumen (*alb.*), whilst on the other there is a certain amount of oleaginous material (*ol.*) opposed to the albumen. Their boundaries, however, are not definitely marked; on the contrary, they insensibly mingle with each other toward the centre of the egg, and it is this indefiniteness that renders the aspect cloudy. Presently the difference in the opposing features becomes stronger and more easily discerned; the albuminous portion grows denser (fig. 16, *alb.*) and more decided in character, whilst the oleaginous substance (*ol.*) assumes a peculiar kind of refraction, totally different from that of the albumen; and in the meanwhile the egg attains to a more clearly defined outline, and grows larger. Soon, now, and while the egg is yet very minute, the albuminous substance (fig. 17, *alb.*) becomes more concentrated toward one side of the egg, and assumes an appreciably definite outline (*p*), but not as yet perfectly globular. At the same time there is initiated within this concreted mass a similar condensation (*w*) at one side. Coincident, usually, with this, the egg, having continued to increase in size, becomes very clearly defined in contour, and the superficial portion develops into a more or less densely accreted envelope, which bears the name of *vitelline sac* (*vs.*). The aim of all these pro-

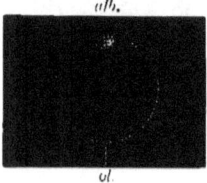
Fig. 15.

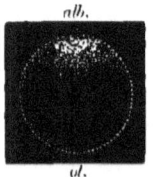
Fig. 16.

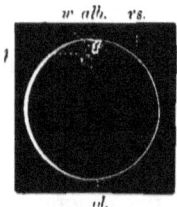
Fig. 17.

Figs. 15, 16, 17, 18. *Theoretical eggs*, representing the process of development from the inception to completion. *alb.* the albuminous pole; *ol.* the oleaginous pole; *vs.* the vitelline or yolk envelope; *p*, the Purkinjean (germinal) vesicle; *w*, the Wagnerian vesicle, or germinal dot. — *Original.*

cesses becomes now rapidly apparent, for soon we find that the albumen has clearly defined itself as a separate mass, (fig. 18,) apart from the yolk (*ol.*), and its superficies has condensed into a well-marked envelope, which constitutes the *germinal vesicle* (*p*), whilst the condensation, going on within it at the last stage, has resulted in the formation of a clearly established agglomeration (*w*) with a distinct wall around it, which is usually called, together with the contents, the *germinal dot*, or sometimes the Wagnerian vesicle. Outside of this field of operations, and antagonistic to it in character, the yolk (*ol.*) has its peculiarities, in physiognomy, refraction, density, opacity, and color, according to the kind of animal in which the egg develops; all tending to demonstrate that it is under a different formative influence from that of the albumen, at an opposite pole, we might say, from the latter.*

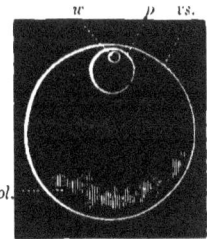

Fig. 18.

We may, therefore, define an egg to be a globular accretion of two kinds of fluids, *albumen* and *oil*, which are always situated at opposite sides or *poles;* but, at the point where they meet, there are various *degrees of separation*, from the most sharply defined line down to the most indefinite boundaries when they are more or less mingled with each other. In the earliest stages of all eggs, these two poles shade off into each other; but whilst some do not develop above this condition, there are others which, as we have seen, take on a higher form of specialization; and in this latter case the concentric vesicles of the egg would seem to bear the character of special organs.

The eggs of all birds, as I have reason to know from almost innumerable observations, possess not only this high degree of specialization in the albuminous region, but exhibit, to an extraor-

* To such an extent is this antagonism, between the oleaginous and albuminous components of the egg, carried out in some of the worms, — *e. g.* Tænia (fig. 44), Planaria (fig. 47), Prostomum (fig. 19), — that the oleaginous portion, or yolk, of the egg is developed in a different part of the body from where the albuminous material originates; and it is not until a certain period that these two are brought into proximity to each other to form a single egg. In this figure

dinary extent, a species of organization in the yolk, such as is observable in no other class of animals. This is as conveniently

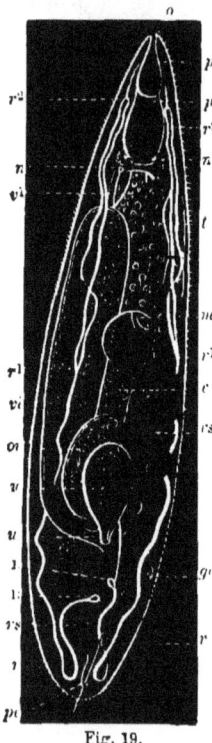

(fig. 19) it may be seen that there are three separate organs necessary to the formation of the egg, namely, the yolk-bearing (vitelligenous) organ ($vi$), which generates the oleaginous substance; the albumen-bearing (germigenous or germ-forming) organ ($ov$), in which the albumen originates and develops into germinal vesicles; and the womb (uterus) ($ut$), into which the germinal vesicles descend and where they are each separately enveloped in a portion of the yolk, which pours into the uterus in certain quantities as it is needed, and there forms, with the germinal vesicle, a perfect egg ($w$). The uterus then forms a shell about the egg, and it is ready to be laid. In other worms, for instance, the *Round-worm* (Ascaris), the formation of the yolk and that of the albumen are carried on in comparative proximity to each other, but yet in different parts of the same organ, and after a while the germinal vesicles (albumen) pass into the yolk (oil) bearing part of the organ, and are there enveloped separately with a quantity of yolk, and the egg is completed. This same process is known to take place in certain Insects; for instance, the common cricket (Acheta), &c. Again, in Spiders, the diversity of local origin is still less than in the last; the germinal vesicle is developed, in a saccular body, to a certain extent, and then the yolk grows about it. Finally, we may find, as in the clams (Venus), &c., the germinal vesicles scattered through a mass of yolk, which in process of time is broken up into groups or concretions, each one of which becomes connected with a germinal vesicle, and surrounded by an envelope which forms a kind of vitelline sac.

Fig. 19. *Prostomum lineare.* (Est. Natural size about 1–1½′′′ (line) long. View of the lower side. $o$, mouth; $ph$, the first part, and $ph^1$, the second part of the throat; $v$, $v^1$, the stomach; $n^1$, the central part of the nervous system, consisting of a large ganglion on each side of the throat, joined by an inferior commissure, $n$; $m^1$, the sucking disc, a sort of prehensile apparatus; $t$, $vs$, $pe$, $ga$, the male portions of this double organization; the female reproductive organs are, $vi$, the yolk-bearing organ, $ov$, the germ-forming organ, and $ut.$ the womb, at present occupied by a single egg, $w$; $rs$, accessory fertilizing receptacle; R, R, the exterior openings of the water canals, $r$, $r$, $r^1$, $r^1$, $r^2$, $r^2$. — *From Schultze.*

demonstrated in the egg of the common domestic fowl as in that of any bird.* If a perfectly fresh egg (fig. 20) is boiled hard and allowed to cool, at least until it may be handled comfortably, and the shell ($s$) and white ($a$, $a^1$) carefully peeled off, the yolk ($ol$) will remain as a perfectly distinct

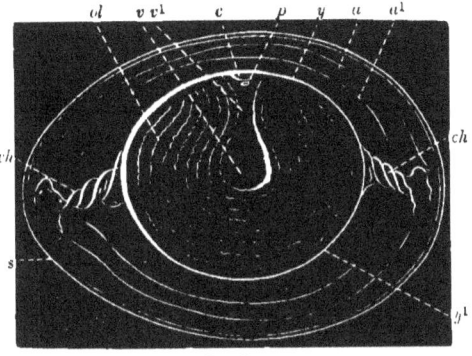

Fig. 20.

Fig. 20. Longitudinal section of a freshly laid *Hen's egg*, which has been boiled. $s$, the shell; $a$, $a^1$, the spirally wound layers of the "white"; $ch$, the innermost layers of the "white" (albumen) twisted into cords, *chalazæ*, which serve as axles, upon which the yolk swings and revolves, whenever the egg is rolled over, so as to keep the side with the white spot, *cicatricula* ($c$), uppermost; $y$, $y^1$, the outline of the yolk mass; $ol$, the concentric layers of yolk; $v$, $v^1$, the more fluid-like part of the yolk; $p$, remnant of the germinal vesicle; $c$, the cicatricula. — *Original*.

* In my investigations of the character of the yolk of Turtles, a group of reptiles which produce eggs most closely allied in character to those of birds, I found an approximation to what occurs in the eggs of the latter. See my observations to this effect in Agassiz's "Contributions to the Natural History of the United States," vol. II. 1857, p. 479, section V.; and note 1, p. 480, note 2, p. 481; also Plate IX., figs. 11–11ʰ, and Plate IXᵈ, fig. 2, with the descriptions of the figures.

*Note on Scientific Property.* It may, perhaps, surprise some of the readers of this volume, upon turning to the "Contributions" mentioned in the above note, not to find my name upon the title-page of the volumes which contain the subjects referred to therein. This is no fault of mine, but of the editor of the series, Professor L. Agassiz. In a little pamphlet entitled, "*A Claim for Scientific Property*," I have published, with somewhat of detail, my claim to a part of the substance of the second, third, and fourth volumes, and a right to have my name upon the title-page of such volumes. It was well understood between Prof. Agassiz and myself, as the result of several conversations, that my share in the investigations should be clearly and distinctly set forth, so that I, in common with him, should be held responsible for the facts, ideas, theories, &c., as far as the language so expressed them; and in accordance with this understand-

spheroidal ball ($y$, $y^1$). It will be noticed, then, that there is on one side of the yolk a light spot ($c$). Taking this spot as a guiding point, and keeping it uppermost, the yolk should be cut into, a short distance, at the side, in such a direction as would divide it into halves, were the knife pressed through so as to strike the middle of the white spot. As the pressure of the knife flattens and smooths over the cut surface so as to obscure

ing, the term, *we*, was used as representative of our combined authorship, and the term, *I*, whenever Prof. Agassiz wished to hold himself alone responsible for anything said. And to keep the distinction always clear, when I myself in turn was the sole authority, my name was mentioned in connection with it, either in a note (vol. III. p. 237; vol. IV. pp. 41, 44), or in the body of the text (vol. IV. pp. 61, 209, 237). In fact, so large a proportion of some of the volumes was the result *purely of my own work* that I ought to have been represented as the *sole authority* where the term, *we*, was used; but as long as I was led to believe that I should share it with Prof. Agassiz, I was content to let it appear so. So clear, indeed, is my own individuality impressed upon these portions which I claim, both in the nature of the microscopic work and the style of the language, a totally different idiosyncrasy from that of Prof. Agassiz's, that, even as much as two years before I had published my " Claim " in the pamphlet above mentioned, European naturalists referred to the investigations as *mine*. This could hardly be otherwise, as every working naturalist knows how such work is done, especially with the microscope; and he is fully aware, too, that it *cannot be delegated to another* to be done for him. The *original authority* must be the one who sits over the microscope day after day, and month after month; and when he takes his pen in hand to describe what he has seen, and what he thinks of it, the very character of the language shows that he who writes has seen and elaborated and thought of what is written; and his fellow-naturalists *will recognize him* in spite of all the efforts of any one who may be so unscrupulous as to attempt to assume the credit through any carefully studied arrangement of the title-page, and the *appearing to give due credit* in an ingenuous preface. Naturalists do not look to the preface to ascertain who is the author, but to the body of the work itself; and any such subterfuge is quickly detected by them. Credit given in a preface, and nowhere else, can only affect and *influence the popular reader, who knows nothing of, and can judge nothing of the merits of the work*. How, and by what representations, I was led to allow this wrong to accumulate in the third and fourth volumes, cannot be entered into here, nor, unfortunately for myself, the facts connected with the matter so easily proved, since the greatly lamented death of one who, could he speak now, might reveal altogether too much for the comfort and effrontery of the one who has done me this great wrong. " *Conscientia insana frons aëneus.*"

what it is desirable to see, the gash should be only sufficient to give a directing tendency, which may be followed up by carefully pulling apart the yolk with the fingers. The structure of the interior will explain for itself why the yolk splits more readily through the white spot than at right angles to that axis. It is impossible, even with the most prolonged boiling, or immersion in spirits, to harden that part of the yolk which extends from the white spot (cicatricula) ($c$) to the centre ($v$) of the spheroid, and consequently there is the least cohesion along that line, and therefore the split takes place most readily in that direction. These same peculiarities, which I have and am about to describe, exist in the egg just before the white and the shell are deposited around it, but the characters are not quite so strongly marked as in the laid egg; and for this reason I have chosen the latter for the better purposes of demonstration, and that you may repeat the observation for yourselves. The principal feature, to which I wish to draw your attention, is the arrangement of the yolk in concentric layers ($ol$), which are alternately dark and light, from the centre to the circumference. These layers, as you will observe, do not form perfectly closed circles, but on the side next the cicatricula bend outwardly and terminate at the circumference. They might be compared to a set of vases placed one within the other, the central one ($v$, $v^1$) containing the uncoagulated fluid which extends from the cicatricula ($c$) to the centre. In the egg, before it has the shell and white deposited around it, the germinal vesicle would then be found just at the mouth of the central vase ($v$, $v^1$); but at this stage there is a mere trace of it, in the form of a clear area ($p$) just beneath the cicatricula ($c$). In specimens of birds' eggs, preserved in spirits, the more highly oleaginous nature of the darker layers is particularly well demonstrated, as the latter assume in that condition a dark orange hue, and the oil oozes out in large quantities at the surface of the fissure, and floats off in drops.

The popular idea of an egg includes the "white," which is merely an accessory, like the shell, and really has no direct relation to the organization of the egg. The concentric, spiral layers ($a$, $a^1$) of the white are the result of several successive deposits in the tube (oviduct) through which the egg passes, with a boring motion, in order to reach the outer world.

Now, as is well known, there are animals which in a full-grown condition are so lowly organized as to correspond in this respect to the early embryonic stages of some of the higher animals, — for instance, a fish in the degree of development of its various organs is comparable to the embryonic state of a horse, sheep, or any of the quadrupeds, — why may we not then have *embryonic eggs* which correspond to the *earliest stage* of the more highly developing eggs? Such, probably, is the state and relations of the eggs (figs. 10, 11) of Balbiani's Spirostomum and Stentor, when compared with those of Laomedea (fig. 14), the Rabbit (fig. 13), and Sow (fig. 12).

From this latter point of view, then, we may look upon the egg as *theoretically* a *bipolar* aggregation of *albuminous* and *oily* substances, and which eventually exhibits a more or less elevated degree of animality; sometimes attaining to an eminent *status*, as when it develops into the most highly organized animals, and in other instances not rising above a very low degree, hardly beyond the egg-stage, properly speaking.

In the latter category Amœba is found, and in fact all Rhizopoda, as you have already been made aware of (p. 9). But let us go on a little further and see how, as we ascend the scale of being, the animal organization proceeds to develop beyond that degree of simplicity which obtains in the egg.

The Sponge is a fair example of those forms which stand, in a transitionary condition, between the Rhizopoda and the next group of animals above them; and we will therefore take it to illustrate the first step in the progress toward a higher state of organization. Although it is difficult to determine whether Sponges are single or compound individuals, it does not affect the question of their relations to the Rhizopoda. That they are undoubtedly above the latter cannot be disputed when we consider that they have a higher degree of specialization, not only in the various functions which they perform, but in the actual differentiation of the various parts of the body. The common sponges of commerce, as they come to us for our daily use, would not help us much to understand their nature; it is only after an attentive study of their living forms, in their native element, water, that one may comprehend the character of those parts of the dried sponge which are not destroyed by the process of preparation to which they are subjected, before they are brought into market. It is a very easy matter to obtain living sponges, because they abound on our rocky coasts, and in our fresh waters. The sponges of commerce, of which there are apparently several species, are not native; nor can any of those which grow about us be used for the purposes to which the foreign sponges are adapted; but yet as far as their nature and general structure are concerned, the native ones are fully adequate to illustrate the group.

The sponge which I have pictured here (fig. 21) is a common occupant of our ponds and streams, most frequently adhering to and growing on the stems of aquatic plants, or forming low prominences on the surface of stones.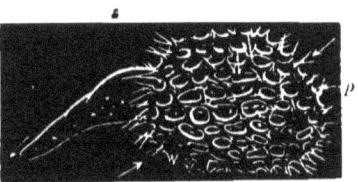

Fig. 21.

It is readily detected by a whitish brown color and bristly surface, and may be found from an almost invisible size to the dimensions of the fist, and in shape either perfectly globular, or oval, or elongated, so as to assume the shape of the stem on which it grows; or, as I have said, as a low prominence, like a bristling wart, on the stones in brooks, or at the side of the pond. It is only when placed in a glass jar and held up to the light, that, with the help of the microscope, its peculiarities are revealed. After the specimen under observation has recovered from the shock of the removal and expanded again, the first thing that strikes the eye is a more

Fig. 21. *Siphydora echinodes*, nov. gen. et sp. 10 diam. A fresh-water sponge. *s*, the emptying conduit; *p*, the superficial interstices, forming a part of the anastomozing channels, and the points of ingress for the water.—*Original.*

or less elongated, finger-shaped, transparent body (*s*) which projects from one side. If this is watched attentively, even with a common pocket-magnifier, it will be found to be a hollow tube; and minute particles may be seen passing in streams through the tube, *outwards*, into the surrounding water. If now, using a much higher magnifier, the surface of the sponge, at the bases of the groups of bristles, be closely examined, there may be detected from time to time the expansion of a minute aperture (*p*), and the passage of particles of matter inwardly. In this way a constant current is kept up, from without to the interior, through these numerous minute apertures; and by a system of variously united canals, the fluid and the included floating matter are transmitted through the body of the sponge and poured into one main channel, and from there they pass into the projecting transparent conduit (*s*), and finally make their exit from its terminal aperture. By this process the sponge obtains its food. The circulation, which is at times fitful, is produced by the vibration of minute cilia which cover the interior of the anastomozing tubes, and by the occasional contraction of the whole body. In the latter case the particles of loose matter within the canals are ejected with considerable violence. The body consists of a soft, highly extensile and contractile, transparent, filmy substance, in which groups of bristles (spiculæ), of a horny nature, are imbedded at irregular intervals, and through which the channels, which I have mentioned, run in every direction. The color is due to numerous granules which lie within its substance; otherwise it would be perfectly glassy in hue; as it really is at the surface, and in the digitiform emptying conduit. Although I have applied the best and highest powers of the microscope to the most transparent and clearest portions of the body, and to the conduit (*s*), I have not been able to make sure that there is a cellular structure in the soft tissue. It is true that occasionally I have detected in the conduit what appeared to be a decided cellular tissue, consisting of distinctly nucleated, closely packed, polygonal cells, but at other times I could not see the least trace of such; nor have other observers been more fortunate than myself; so that in this respect sponges have as low a type of tissues as the Rhizopods.

But yet you must have seen, that, as I said at the beginning of this description, the sponges have a degree of specialization, of parts of the organization, that at once stamps them as of a higher order of beings than the Amœbas and their congeners. In addition to what may be found in the Rhizopods, we have in the Sponges distinct channels of circulation, vibratile cilia giving a direction to these currents, particular inlets and as definite an outlet for the passage of food; and in some of the peculiar forms, the spiculæ are arranged according to definite patterns about the inlets and the outlet, thus adding an element of gradation, while subserving the needs of an already distinctly specialized function.

The group of beings to which the next animal, that I shall introduce to your notice, belongs, is one that in a certain sense is more directly allied to Rhizopods than to Sponges, on account of the mode of life which is prevalent among the members. But what distinguishes the Actinophryians, as this group is called, from the two foregoing is that they, at least some of them, have a distinct cellular structure; and yet, though the cells are very distinct, they exhibit a low state of development, as low perhaps as could possibly obtain without failing to be genuine cells. In other respects the nature of the organization is very much like that of Rhizopods. Still, from another point of view, the Actinophryians might be called a peculiar group of Sponges; in fact, there are some forms among them that are as yet in an undecided position, in the opinions of naturalists, as to whether they are members of the latter or the former group. The truth of the matter is, they form a transition from the one to the other, of such insensible gradations that it is impossible to determine where the one group ends and the other begins. It is on this account, therefore, that out of these innumerable gradations I can only present for your consideration those forms, in the progressing scale of development, which exhibit such features as mark a distinct step in the elevation of animal organization.

This figure (fig. 22) represents the internal structure of Acti-

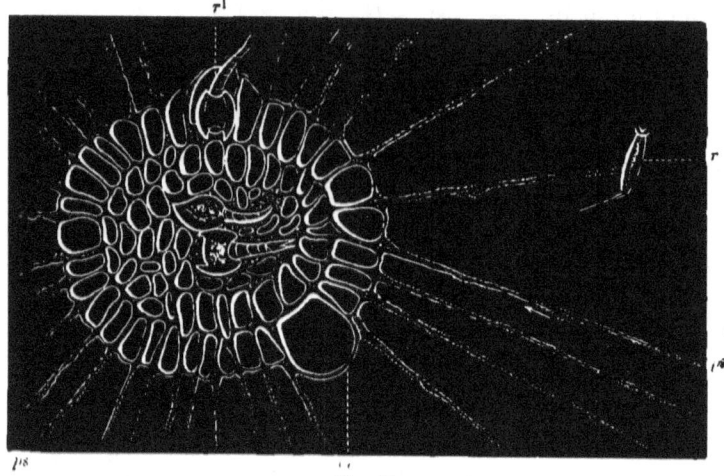

Fig. 22.

nophrys, the typical genus of this group. It abounds in still waters and ditches, from whence it may be collected and preserved alive by skimming the surface of the soft silt, and depositing it, for observation, in a glass jar of clear water taken from the same spot. Even to the naked eye it is visible as a distinct, white, glistening spot, apparently about as big as a pin-hole. Under a moderate magnifier it resembles a small sun with greatly prolonged rays. It is on this account that one of the species has been called Sol. Although, as we shall see presently, capable of moving parts of its locomotive apparatus with great rapidity, it progresses at a very slow pace, balancing itself on the tips of its rays ($ps$), which project from all sides of the body. In fact, it is so sluggish that it may be handled, with a great deal of freedom, without inducing it to retract its *pseudopodia* ($ps$), as its rays are called. Notwithstanding that the pseudopodia are capable of being totally contracted, so as not to leave a trace of

Fig. 22. *Actinophrys Eichornii.* Ehr. 130 diam. A view of the interior, as if it had been cut open at the middle. $ps$, the pseudopodia acting as organs of locomotion and for prehension; $r$, living prey, a Rotifer, just caught; $r^1$, another Rotifer in the process of engulfment; $cv$, the contractile vesicle, one of the cells of the outer layer. — *Original.*

their presence, they possess a power not only of extending to a great length, but also of rendering themselves so rigid that a few of them can sustain the whole weight of the body. As a general thing they project from the body in perfectly straight lines, and move so slowly as to appear like stiff bristles, rather than what they really are.

By a close and patient examination, however, one is not long in coming to a conclusion that these apparently sluggish creatures are not only animals, but at times very active and powerful. These particulars are better understood when the intimate structure of the Actinophrys is under consideration than by any other process of observation. We will therefore proceed to investigate the basis of its organization, as it is exhibited in its cellular tissues. The most comprehensive view of its structure, that it is desirable to take, is that in which we get what is called a sectional view, as if a section of the body, next the eye, had been cut away and the interior exposed to the sight. This is done by placing the glasses of the microscope so as to get a view of the interior, whilst all those parts nearer to the eye are out of focus, and consequently not defined so as to form a distinct picture. The figure before us is a sectional view, so deep that the body is as it were cut into halves. By it we get an insight of the relations of the cells to each other, and to the pseudopodia; in fact every part of the organization may be seen at a glance. Under a low magnifying power the body appears as if it were a mere globule of gum-like or mucous matter, with numerous closely set cavities hollowed in its substance; but a closer examination, with good lenses of a higher power, will detect a distinct cell-wall about each cavity. Where the cells lie close together the neighboring walls appear as one; but at other points, which are numerous, the cells are separated by the interstitial mucous substance, (*cytoblastema*, cell-generator,) and there the wall may be seen to have a very appreciable thickness. This is a very important point to determine, because by this feature alone the Actinophryans are to be estimated as more highly organized than those groups, the Rhizopods and Sponges, which possess a mere mucous tissue. In the latter case we see the lowest possible grade of organic tissue, whilst in the former it is specialized so as to present two distinct forms, namely, the mucous form, (cytoblastema,) and the cells which have been generated in it.

At once, then, you will see that the organic functions are distributed not only in different regions of the body, — as we have already observed among the successively rising members of the Rhizopod group, — but among two distinct sets of tissues. In the former groups everything is performed by a body that is all cytoblastema, but in this group the cytoblastema has generated cells which assist in performing the functions of the organization. Moreover the cells are differentiated so as to present two distinct features among themselves; thus we have an outer layer of cells ($ce$, $r^1$), which are much larger than those within, and they are disposed quite methodically in a single layer all over the body; and within, the smaller cells are united, without apparent regularity, so as to form a sort of core. In the interstices of both kinds of cells there is a universally pervading cytoblastema, which also at certain points has a specialization of its own substance. I refer to its prolongation, from between the cells of the outer layer, into those attenuated, bristling bodies which I have spoken of as pseudopodia ($ps$).

It is a notable fact that although the cytoblastematous substance overlies the whole body, exterior to the larger cells, yet it never projects, in the form of pseudopodia, as if in prolongation of the cells, but invariably alternately with them.

Notwithstanding the simple structure, or rather structureless character, of the pseudopodia, they at times exhibit a rapidity of motion equal to that of the most highly organized muscle. This is most frequently seen in connection with the seizure of living animals for food. Minute creatures of almost every kind are a prey to this far-reaching Briareus. The moment that any moving body comes in contact with one of the pseudopodia, — as for instance the little Rotifer, a shrimp-like animal, which I have represented here ($r$), — it becomes as it were glued to it, and stupefied; the pseudopodium then either gradually retracts, or, as I have frequently seen, suddenly, and as if it were a rubber thread under high tension abruptly cut loose at one end, jerks the prey toward the body. At the same time the neighboring pseudopodia bend over the victim and form a sort of cavity ($r^1$) about it, and then the surface of the body rises on each side and gradually engulfs the living morsel, which by this time shows but little signs of life. In a short period it is passed through the cortical layer of cells to the interior, and there undergoes digestion in a fluid which seems to be generated, in a cavity between the cells, for each special body that is introduced.

As yet we have not arrived among those animals which possess a distinct mouth, but the food is introduced at any point of the body which corresponds to the interspaces of the cells; and likewise when the digestible parts of the prey are extracted, the refuse is ejected at any spot in the circumference. There is also a tendency toward a circulation of fluid, most especially exhibited in the pseudopodia; but it is more apparent than real, and requires the most careful scrutiny to detect the actual state of things. The minute transparent granules that more or less abound in the cytoblastema, and particularly in the pseudopodial prolongations ($ps$), are the only means by which we are enabled to descry the movement of this tissue at any particular point; and as they are borne along in dilations, extensions, or retractions of the various regions of this glairy substance, they appear to be floating in streams of fluid, whereas their motions are really limited to the extent of the expansibility or contractility of the cytoblastema.

There is, however, in certain parts of the body, a more reliable exhibition of an incipient fluid circulation. Among the large cells of the cortical layer, one or two of them ($cc$) exhibit a tolerably regular but slow expansion and dilatation, like the diastole and systole of the so-called heart, or *contractile vesicle*, of the higher Infusoria, which I shall speak of presently. In what manner these vesicles are connected with the clear fluid with which they fill themselves, when they expand, or through what channels they expel it, when they contract, has not been discovered; although in all probability it is strained through the cell-wall.

Such are a few points of the structure of this marvellous creature. For hours I have watched it with an ever increasing interest, and when I left my microscope I could only feel that but a beginning had been made in the investigation of its varied functions. Indeed, I hardly know where to stop in the enumeration of the rapidly accumulating differentiations of organs and the specializations of functions. We have scarcely arrived in the midst of creatures which possess barely such a sufficiency of structure of an organic nature as would enable us to distinguish them from inorganic bodies, before we light upon, I might almost literally say, numerously appointed functions, each devoted to a separate work. This, no doubt, teaches us that we are not to look to any peculiar and absolute form of combination of the chemical elements, Carbon, Hydrogen, Oxygen, Nitrogen, by which we may distinguish the organic from the inorganic kingdom of nature. The variously performed functions of the extremely simple Amœba gave us the hint toward this conclusion; and now the but little more highly organized Actinophrys more than redoubles the impression upon our minds, that it is a power or force, or, as I have stated in the beginning of these lectures, *a principle of life*, — whatever that as yet uncomprehended principle may be, — that constitutes the *vitality* of the *organic being*, and distinguishes it from the *inorganic thing*.

The most highly organized creature, even man, is no more an animal, as distinguished from the mineral, than Amœba is; for no one will pretend to say that when he existed in that embryonic condition, the egg-state, which is as simple in structure as the Amœba, that he was any the less *animate*, that he was any the less an *organic* being, than when in an adult state. The more highly elevated and the infinitely more numerous functions of man are not any addition to, but merely so many *variables* of, the simple principle, *vitality;* alike as potential to vivify the Amœba as the Man; but no further removed from the principle that rules the mineral, in the most highly complicated, than in the most lowly organized beings.

Doing away, then, with the idea of the necessity of a more or less complicated structure for a medium in which to exhibit the principle of vitality, as distinguished from the *inanimate*, we can readily imagine that a living being could be possible, were it no more complicated than a drop of water; and when we come to this conclusion, we have but to call to mind the initiatory stages of the animal, in its *egg-form*, to realize this thought in its perfection.

The most infinitesimal drop of fluid, that is potentially an egg, is as truly so as the most complicated, and likewise as certainly an animate being, as on the day that commences its career as a wanderer over the earth, whether in the waters as a fish, or on land as a quadruped, or in an erect position, a Man. Man or Monad, the mighty oak or the slimy mould of our cellars, are alike the medium for the exhibition of the principle of vitality; nor can we say how simple that body ought to be which might not be subjected to the dominion of this power.

But let us return from this digression, and see, if possible, whither the group of Actinophryians will lead us, if we pursue the investigation of the successively higher forms. I will stop but a moment to point out two or three of the characteristic features of one of the Polycystinæ, (fig. 23,) a very peculiar group of animals, which stands related to the three groups which I have already discussed; to the Rhizopods, like Cornuspira (fig. 3) and Rotalia (fig. 4), it is allied by its thread-form pseudopodia (fig. 23, *ps*), which are projected through the apertures of a shell; to the Sponges

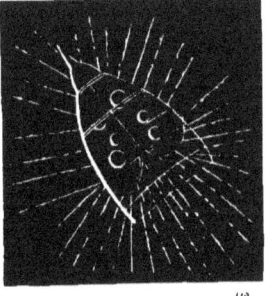

Fig. 23.

Fig. 23. *Lithocampe tropeziana.* J. Müller. Nat. size. $\tfrac{1}{15}'''$ (line) long. A mitre-shaped shell enclosing an Actinophrys-like body. Marine. *ps*, the pseudopodia projecting in every direction through the pores of the shell. — *From J. Müller.*

it is more distantly related through a group of sponge-like creatures called Acanthometræ, which combine in their organization spicules, resembling those of Sponges, and a perforated net-work-like shell, similar in conformation to that of many of the Polycystinæ; to the Actinophryians (fig. 22) it so closely approximates that it might be called an Actinophrys, with a stony network thrown over it.

The most direct line, by which we may pass from Actinophrys to the higher Infusoria, is through the mediation of a very singular creature which was discovered, by Dr. Strethill Wright, on the coast of Scotland, near Edinburgh, and to which he gave the name of *Zoöteira religata* (fig. 24). It is, as he says, "an Actinophrys mounted on a contractile pedicel." At times its pseudopodia (*ps*) are extended into extremely attenuated threads, and at others they are "all thickened or clubbed at their extremities." This figure (fig. 24) represents them in the latter condition, and the tubular stem (*s*) so expanded as to render its net-like character (*n n*¹) very conspicuous. The axis of the stem is occupied by a "muscular band" (*m*), along the centre of which fluid was seen to circulate on one occasion. The whole animal can retract itself into the gelatinous sheath (*sh*) which surrounds its base. Its mode of catching its prey and engulfing it is precisely like that of Actinophrys.

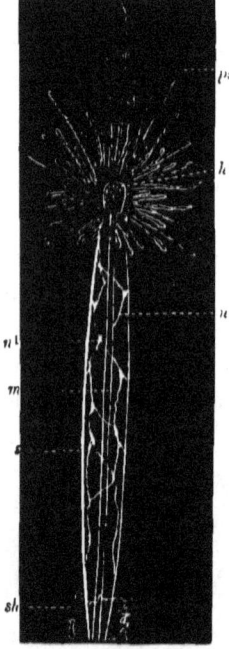

Fig. 24.

The next animal that I shall draw your attention to, although it is not directly and closely related to Zoöteira, is, however, in

Fig. 24. *Zoöteira religata.* Strth. Wright. Magnified considerably. A stalked Actinophryian. *h*, the head; *ps*, pseudopodia; *s*, stem; *n*, *n*¹, net-like threads of the interior of the stem; *m*, the hollow muscular band in the axis; *sh*, the sheath. — *After Strth. Wright.*

some respects similar to it, and stands in the course of the transitions toward the highest Infusoria. It is called Podophrya (fig. 25). I introduce it here principally because, in its adult state, its character and habits are strikingly like those of some of the Rhizopods; for instance, it seizes its prey with its globe-tipped feelers ($f$), and through them sucks the juices of the victim. The same we have seen is done by some of the Rhizopods which live in a perforated shell. (p. 14.) In its embryonic state (fig. 25, A, B) it is evidently a close ally to those higher Infusoria which move by vibrating cilia, and, like them, the young (B) exhibit in the arrangement of the cilia an obliquity which points to the spiral type of conformation, which, as I shall show hereafter, is at the base of the organization of Protozoa. What chiefly gives it a rank above those Protozoa which we have already taken note of, is the definite character of its contractile vesicle ($cv$), and the reproductive organ ($n$); both of which can scarcely, if at all, be distinguished from the corresponding organs of the highest Infusoria. In Podophrya, the contractile vesicle ($cv$) is readily distinguishable from the rest of the tissues; nor is it, as in Actinophrys, likely to be mistaken for one of the ordinary cells of the body, as it is differentiated in such a manner that its physiognomy is peculiar to itself, and it seems like an isolated cavity in the midst of the animal. With regular precision it slowly contracts to an almost invisible point, and then expands to its former rotundity, and again and again repeats the systole and diastole, with ever-recurring, evenly marked intervals.

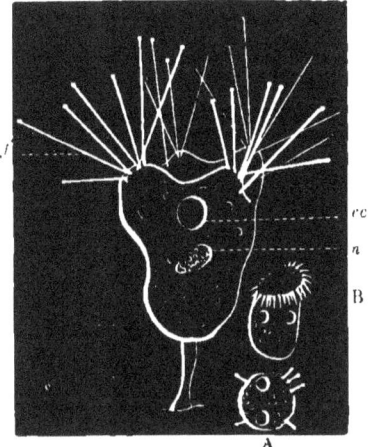

Fig. 25.

Fig. 25. *Podophrya Cyclopum*. Clap. 300-350 diam. $f$, the globe-tipped feelers; $cv$, contractile vesicle; $n$, the reproductive organ. A, B, the young in different stages of development. — *After Claparède.*

And so I might go on upwards, from one kind of creature to another, showing point by point the gradual increase in the complication of animal organization, until I arrived at the highest forms, even at Man himself, who, of all animals, departs most from that degree of simplicity which we find in the egg. But it will be more proper to trace these gradations, in full, in another part of my subject. It does not necessarily follow here, from what I have said, that there is a serial relation from the lowest to the highest animals; that is another thing. I simply mean to assert that the various degrees of complication are not suddenly marked off, with gaps between them, but that the idea of differentiation has been carried out gradually. My present object is merely to impress upon your minds the fact that there is no sudden transition from the *condition of an egg* to that of *an animal*, taken in its usual sense.

Being thus preoccupied with the idea that there are animals (Amœba, Actinophrys, etc.) as simple as some eggs, or even more simple than others, you are prepared for the assertion that an egg is not to be looked upon as a distinct *body*, which *preëxists* the animal, but rather that it is *the animal itself*, from the moment when it begins to form in the ovary of its parent. The egg is merely the *first stage of growth of an animal*, and it is not separated from the succeeding phases, any more than these latter are from each other.

From this, and what has already been told you preceding this, you may draw the inference that there is a perfect parallelism between the development of an animal from the earliest or egg-stage to the adult, and the successive degrees of grade from the lowest to the highest animals, within a group.

Now, in regard to the point of origin of the egg, the fact that it is formed *within a parent* rather than in the outer world is perhaps only a *difference of degree;* for although some eggs are retained by the parent until after the egg-stage is passed, in fact until the time when the young is able to move about and take care of itself when born, as in our common quadrupeds, yet even here there is a marked difference among the successively

## OF THE GROWTH OF ANIMALS. 53

lower ranks, and a more or less corresponding decrease in the degree of dependence of the young upon the parent; for instance, in the duck-billed quadruped, or Ornithorhynchus, (fig. 26,) the young are born while yet in a far inferior state of development to that of the young of the horse, or cow, or dog, when born. The young of the kangaroos and opossums, of which there are many species, are hardly more advanced when born than the young Duck-mole. But there are vertebrates which in a descending scale are for a less and less time retained; the eggs of birds are laid, that is, subjected to the influence of surrounding external causes, at a time when there is scarcely a trace within them of the so-called *germ*, which from this period is altogether dependent upon one of the physical agencies, *heat*, for its development. Still lower in the scale we find the eggs of frogs and toads, and some kinds of fishes, are laid before they can hardly be said to have become fully formed *as eggs*. And so we might go on, pointing out instances of this decreasing degree of dependence, until we find the egg expelled from the parent at such an early stage as to be dependent upon *physical agents during a large part* of its period of growth; for instance, it is dependent upon temperature, or moisture, or dryness, or light, or some form of *physical help*, exactly as are spontaneously generated bodies, or as those developed in the sealed flasks.

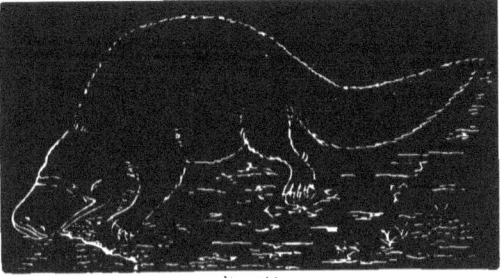

Fig. 26.

Fig. 26. *Ornithorhynchus paradoxus.* Blum. Natural size, "as large as a cat." The Duck-bill quadruped, or Duck-mole. Inhabits Australia. — *After Bennet.*

## CHAPTER III.

THE OLD APHORISM, "OMNE VIVUM EX OVO," NOT STRICTLY CORRECT. — THE ORIGIN OF INDIVIDUALS BY BUDDING AND SELF-DIVISION.

HAVING thus shown to you that there are among adult animals those which are as simple as the lowest forms of eggs, and that eggs are merely one stage, and that the earliest, of animal life; and having demonstrated that the so-called egg is left more or less dependent upon physical causes for its growth, in some cases almost altogether independent of the influence of the parent, and consequently in like proportion dependent upon physical agency for its growth, there seems to be but one step left to bring us to that condition of things under which the animal-egg may arise *altogether independent of a parent!*

This idea may seem at first startling and unnatural, so accustomed are we to look upon all animals as direct developments from maternal parents. Yet there are numerous instances, well known, and acknowledged as such, by all naturalists, of individuals which originate in such a way that they may be truly said never to have been born; that is, they have never passed through or existed in an egg-state or condition.

Now, in order that this may not seem to be an incidental assertion, but that it shall appear to you in its true light, which is, that an individual is not always derived directly from a preliminary egg-phase, I shall illustrate the phenomena by a pretty full description of the various modes of reproduction otherwise than those which take place through the means of ovarian gestation.

There are, in the first place, two apparently well-marked kinds of individuals which originate by budding: the one is an individual in the truest sense, a complete, independent organization, and the other is as fully complete in all its parts, but at the same time lives in common with others like itself, and forms what is called a compound individual. Between these two kinds there are, however, all possible gradations; but as it is not germane to my purpose to describe them, I merely mention the fact, for the sake of future reference, and then pass on to what is most pertinent to our subject.

The strictly independent individual is one in which all the parts of the organization, which belong to animals of this or that particular group, are fully represented in a single body. To commence with a most familiar instance, I will draw your attention to the animal which is represented here (fig. 27). It is known by the name of Hydra, and belongs to the same type of animals, namely, the Zoöphytes, as the jelly-fishes, corals, starfishes, &c., but is one of the simplest of them all, in every respect. It is to all intents and purposes a simple elongated sac (*s*), with slender, hollow prolongations (*t*) arranged around its mouth. These prolongations, which vary in number from five to eight, are called the tentacles, and are used as feelers, and for the purpose of seizing the food, which is mostly living animals, and conveying it to the mouth.

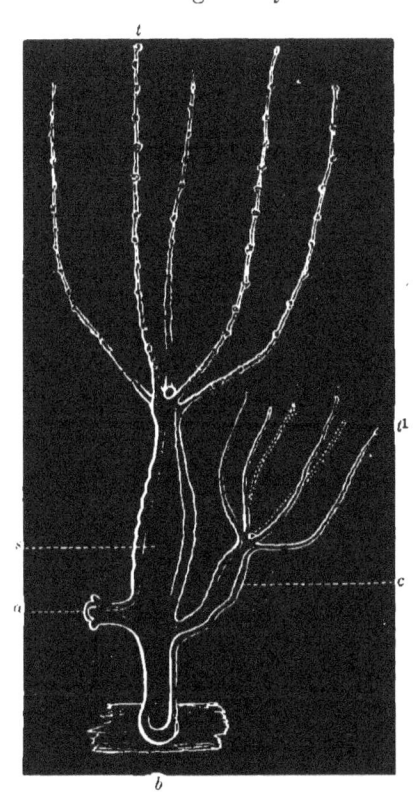

Fig. 27.

The latter opens at the end of the

Fig. 27. *Hydra fusca*. Trembly. 14 diam. The Fresh-water Hydra, with two young (*a*, *c*) budding from it. *b*, the base, attached to a piece of stick; *s*, the digestive cavity; *t*, tentacles of the adult; *t¹*, tentacles of the young. — *Original*.

little conical eminence at the base of the prehensile organs (*t*). The opposite end (*b*) of the body is closed, and slightly expanded in the form of a disc, by which it attaches itself to various objects, such as pond-lilies, duck-weeds, or even to the sides of stones on the margins of lakes. At certain seasons very few individuals are to be found which are not in the condition which we have represented here, that is, having either one, two, or three younger individuals (*a*, *c*) growing out from each single adult. The process of this kind of reproduction is very clear: the bud begins in the form of a simple bulging from the side of the sac-like body; it increases by a mere prolongation, as if the wall were puffed out in the form of a hollow cylinder with a rounded end, and presently minute processes rise around this end, and produce a form such as stands out from the body on our left (*a*).

We have, now, all that is essential to the new individual, but in a rudimentary condition. To attain to perfection, then, the cylinder elongates, as represented on our right, (*c*,) and the minute processes, the tentacles, simply develop into thread-like bodies, ($t^1$,) whilst the rounded end becomes more prominent and conical, and a perforation appearing therein, a mouth is formed. In this condition the little creature is prepared to seek its own prey. Its independence is finally accomplished by a gradual constriction of the base of the new body, at the point where it is attached to the old stock, until it finally, as it were, cuts itself off. From this time, it is as purely an individual as the one from which it budded, and, like that, it reproduces its own likeness, and apparently without limit as to numbers. Sometimes this occurs before the young stock is detached from the primary one, and in this way a numerous colony is produced, which has all the physiognomy of a minute branching water-weed. To heighten the deception, some species of Hydra are green, and in their most expanded state, with the tentacles spun out to excessively fine threads, they resemble tufts of green silk waving backwards and forwards under the influence of varying currents. Finally, however, the whole ramified mass scatters into numerous branchlets, and each individual pursues its own independent course.

Now it is remarkable that the proportion of individuals which are produced in this way is so great, when compared with the number of those which arise directly from eggs; and one might almost say that the latter process is the exceptional one, and the former the normal mode of reproduction.

Another instance of the budding of independent individuals among the Zoöphyta is exemplified in the common Sea-Anemone, or Animal-flower, so-called. This is a much more highly organized animal than Hydra, although it belongs to a group which, as a whole, is more simple in structure than that of which Hydra is a member. The latter is classed among the jelly-fishes, "sea-blubbers," "sting-bladders," &c., but the former is one of the Coral group. The body of the Sea-Anemone, which we have represented here (fig. 28), under the name of *Metridium*

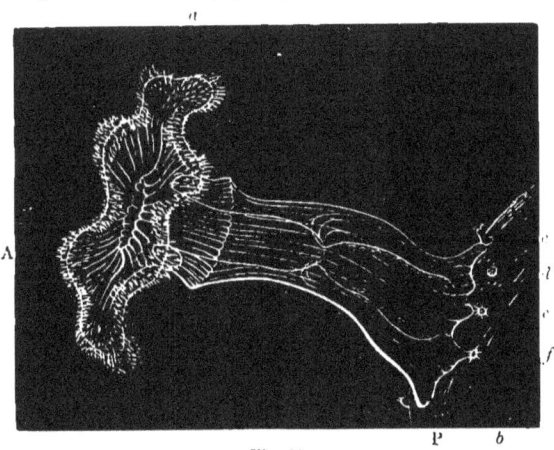

Fig. 28.

*marginatum*, has the appearance of a cylinder cut straight across at each end, and at both of these points covered by a membrane. One end (A) is bordered by an undulating fringe (*a*) of numerous short finger-shaped tentacles, and the other extreme (P) is more or less broadened like the base of a pillar,

Fig. 28. *Metridium marginatum.* M. Edw. ½ natural size. A Sea-Anemone attached to the shell (*b*) of a mussel. From Boston harbor. A, the anterior end; P, the posterior end; *a*, the fringed disc, covered by pointed feelers, and pierced by the scalloped, oblong mouth; *c, d, e, f,* young budding. — *Original.*

and forms the means of attachment. At the centre of the membrane of the fringed free end (A) of the body is an elongated opening with a scalloped outline. This is the mouth. From this

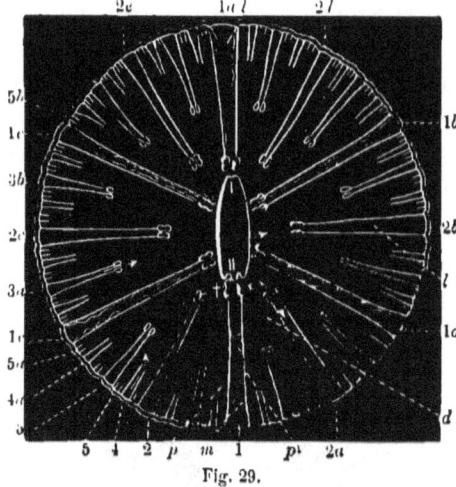
Fig. 29.

a flat sac (fig. 29, I, II,) projects as far as half-way to the bottom of the cavity of the cylinder, and each side of this stomach, as it is called, corresponds to the two elongated sides of the mouth. At its end there is an opening which leads into the space which surrounds it, the so-called digestive cavity ($d$).

The latter is divided into numerous chambers, ($l$, $d$,) extending from one end of the cylinder to the other, by thin semi-partitions ($p$, $p^1$) which, at one edge, are attached, always in close-set pairs (numbers 1, 1$a$, &c., to 5, 5$a$, &c.) for their whole length, to the interior of the cylinder; at each end they are united to the transverse membrane, or, as they are called, respectively, the floor of the oral disc (fig. 28, A,) and the floor of the base (P); and lastly, along their edges, which are turned toward the centre,

Fig. 29. *Cereus Sol.* Verrill. 2 diam. From Charleston, S. C. A fore-shortened view of the interior, from just behind the mouth, backwards to a point a little behind the posterior end of the stomach (I, II). $d$, the general digestive cavity; $l$, spaces between the pairs of partitions, the arrows indicating the passage-ways to and from the general cavity ($d$); $p, p^1$, the two partitions of one of the largest pairs; $m$, the muscular layer on the opposite faces of $p, p^1$; *, †, the reproductive organs, which in profile (fig. 28) appear like a deep frill along the edges of the partitions, extending from the posterior end (fig. 28, P) of the body toward the head; I, II, the two plicated, upper and lower, edges of the flat stomach; 1, 1$a$, 1$b$, 1$c$, 1$d$, 1$e$, the six pairs of the first set of partitions; 2, 2$a$, 2$b$, 2$c$, 2$d$, 2$e$, the six pairs of the second set; 3, 3$a$, 3$b$, &c., the twelve pairs of the third set; 4, 4$a$, &c., the twenty-four pairs of the fourth set; 5, 5$a$, 5$b$, &c., the forty-eight pairs of the fifth set.—*Original.*

a part of them join the stomach-wall as far as it extends, and from the entrance to the digestive cavity (fig. 29, *d*) to its bottom, they all have a free edge (fig. 29, *, †).

There are never less than twelve of these semi-partitions, not even in the youngest; and when more numerous they occur in multiples of six. Thus, in the youngest individuals, we have six pairs (1, 1*a*, 1*b*, 1*c*, 1*d*, 1*e*) of twelve, which are attached, two (1) at one edge (II), and two (1*a*) at the other edge (I), of the flat stomach, and the four remaining couples, (1*b*, 1*c*, 1*d*, 1*e*,) dividing the area on each side of this organ into three equally broad spaces, (namely, from 1 to 1*d*, from 1*d* to 1*b*, and from 1*b* to 1*a*,) are joined to it, as at its edges, (I, II,) from the mouth to its opposite opening. From the latter point, all six pairs have free edges, and divide the general cavity, behind the stomach, into six principal broad spaces, or berths, as we might call them, and six narrow berths, which are, respectively, the spaces (*l*) between the two partitions of each pair. In this way each half of the body, on the right and on the left of the flat of the stomach, is divided into three broad and two narrow longitudinal spaces. In older individuals, each of the broader spaces is divided again by a double partition, (2, 2*a*, 2*b*, 2*c*, 2*d*, 2*e*,) and in still more advanced ones, each of these twelve are halved in the same way, (3, 3*a*, 3*b*, &c.,) and so on until in very old specimens there are, as I have counted, as many as sixteen times six of these broader berths (*d*) and as many of the narrower ones (1 to 5). Each of the successively formed sets of berths are, in regular progression, less deep than those which originated before them, and, in like manner, the more recently formed partitions are successively narrower (numbers 1, 1*a*, &c., to 5, 5*a*, &c.) than the older ones.

The tentacles, which are hollow, in every instance are placed so as to overlie each narrow space (*l*) between a pair of partitions, and form, as it were, saccular prolongations of these spaces. In the older individuals, the edge of the oral disc is more or less wavy, and the hollow tentacles which cover it form the fringe (fig. 28, *a*).

The reproductive organs, which consist of wavy, narrow bands, are attached to the free edges of the older sets of semi-partitions, and correspond in size to the organs to which they are united; the largest (fig. 29, *, †) bordering the oldest partitions, and the successively smaller ones occupying the edges of the correspondingly lesser partitions (2 and 3). You will see by this, that, as a matter of course, these organs occur in pairs, just as do the semi-partitions, to which they form a border.

Thus far the description of the animal was necessary to an understanding of the nature of its organization; but as for the other details I shall leave them to be taken up when considering it from another point of view.

The budding always takes place from the end by which the Anemone attaches itself, and, as a rule, at the edge of the so-called basal disc. In the course of a month I have seen as many as twenty buds detach themselves from the parent, and removing to a short distance, they resembled a group of attendants (*c, d, e, f*) surrounding their master. As in Hydra they arise as simple rounded protuberances, but in a short time six short tentacles make their appearance at the free end, and a minute oblong aperture, the mouth, is formed in their midst in such a way that its two ends have a tentacle opposite * each, and the other four disposed, two on one side and two on the other.†

Within, the organs arise at points corresponding to the position of those outside. The semi-partitions, twelve in number, begin as mere ridges, which extend in pairs from the anterior end of

---

* See the description of the young Anemone at the beginning of Chapter X. for more complete details in addition to what is given here.

† It is a remarkable fact, that, in these budding young, the tentacles are very rarely if ever perfectly symmetrical on each side of the elongated mouth; and so it is also with the internal organization, *i. e.*, the semi-partitions and reproductive organs. In this respect the single Polyps correspond to the compound branching corals, in which the buds, never becoming detached, but remaining through life, have a more or less one-sided attachment. Yet this asymmetry of the successively arising heads, along the growing branch of a coral, would seem to have a direct relation to their spiral arrangement, precisely in the same way that the asymmetry of the leaves of a plant, the petals and sepals of a flower, and the scales of a Pine-cone, has a direct reference to the spiral in which they are disposed.

the stomach along the oral wall toward its border; but as they are not conspicuous, the general cavity of the body appears as simple as that of Hydra. Eventually, however, these organs broaden, and extending along the inner face of the cylindrical body toward its base, produce the berths which I spoke of just now when describing the adult. The period at which the young are detached from the parent varies to a considerable extent, but usually it occurs before the tentacles have more than doubled their number.

Sometimes several pieces of the base of an Anemone separate from it before there is the least trace of tentacles to be seen, and as in this case they do not at first show any signs of activity, but on the contrary remain for a long time in a quiet state, they have the same appearance as artificially separated pieces, and like them seem to be undergoing a recuperative process, after the shock of separation. On this account these voluntarily separated pieces arise into perfect individuals, after being detached from the original stock by what is called *voluntary self-division.*

Occasionally an adult Anemone has been seen to divide itself longitudinally into two equal parts; but this does not occur as a rule except among compound Polyps, (Corals,) and in that case there is not a complete separation of the resultant individuals, but only the heads and stomachs are divided, whilst the general cavity remains common to the two.

Before proceeding any further I may as well anticipate what you would soon observe, and that is, that there is no well-marked difference between budding and self-division. This you may have suspected from a comparison of the modes of budding and self-division of the Anemone; but in the example, which I will now introduce, the two processes are undoubtedly carried on at once in the same individual.

In all *Protozoa* the process of self-division, as it is usually, but mistakenly called, occurs as one of the normal methods of propagating new individuals; and until of late years it was thought to be the only one. In the works of the older authors these animals were described as simply cutting themselves in two by a kind of constriction, but in later years observers have discovered that the process is attended with considerable preparation, and from this it came to be known that it is, as I have said, a compromise between self-division and budding.

In order that you may fully appreciate the character of this compromise, I must first illustrate some parts of the organization of the Infusorian which I have chosen for this purpose. It goes by the name of *Trumpet-animalcule*, (fig. 30,) on account of its

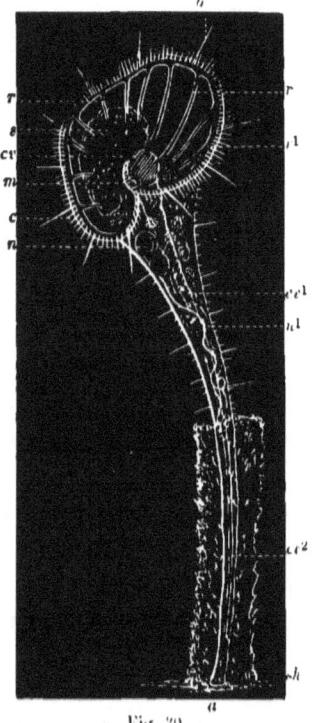

Fig. 30.

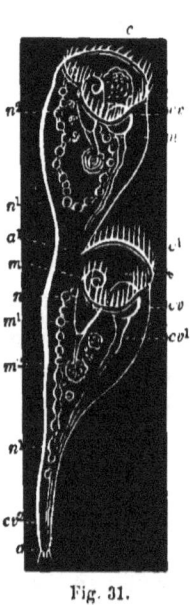

Fig. 31.

Fig. 30. *Stentor polymorphus*, Ehr. 130 diam. The Trumpet-animalcule, expanded and bent slightly over toward the observer; the mouth ($m$) next the eye, and the dorsal edge in the distance. $a$, the posterior end; $sh$, the tube enclosing $a$; $m$, the mouth; $c$, the ciliated border of the disc ($s$); $b$, the larger, rigid cilia; $cv$, the contractile vesicle in the extreme distance, seen through the whole thickness of the body; $cv^1$, $cv^2$, the posterior prolongation of $cv$, in the distance; $r$, $r^1$, the circular and radiating branches of the nervous system; $n$, $n^1$, the reproductive system, extending from the right side, at $n$, posteriorly, but toward the eye at $n^1$. — *Original*.

shape. Its scientific name is Stentor. It is very common in ponds and ditches, forming more or less extended colonies on the surface of the stems of water-weeds, or submerged sticks and stones. Some of the varieties, for they can hardly be called species, have a deep cœrulean blue color, and on this account their presence is readily detected by the sudden shrinking of blue patches of what appears to be a sort of blue mould, when they are touched, or the stem to which they are attached is shaken. The fully expanded variety, which I have figured here, is perfectly colorless, and constructs a distinct tube (*sh*) of gelatinous matter, into which it retreats, when disturbed, and shrinks into a globular mass. Its narrower end (*a*) is attached to the bottom of its sheath at the will of the animal; and although it seems rarely to leave its domicile, yet when it does so it has the power to detach itself without violence, and move away to any other spot, and there affix itself again and build up a new tube. The most striking feature about this animal is the one-sidedness of its figure, as if the edge of the mouth of the trumpet had been cut off obliquely, and the rim stretched out at one point. On the side next the observer's eye the edge turns inward (at *m*) so

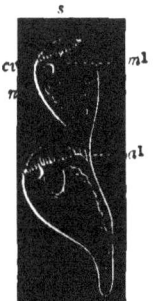

Fig. 32.    Fig. 33.

Fig. 31. *Stentor polymorphus*. Ehr. 100 diam. A much larger specimen than fig. 30, from the same point of view, undergoing the process of fissigemmation, whilst swimming about in a half expanded state. *c*, the head or disc of the original stock; $a^1$, the point of separation, which is at the tail of the old stock and the right side of the head $c^1$ of the new individual; *m*, mouth; $m^1$, throat; $m^2$, posterior end of the throat; *c*, ciliated edge of the disc; $c^1$, the as yet incomplete ciliated edge of the disc; *cv*, the contractile vesicle in the distance; $cv^1$, $cv^2$, the posterior tubular prolongation of *cv*; *n*, $n^1$, $n^2$, the reproductive organ, more or less swollen from point to point by the enlarging eggs; *s*, the disc of the new individual, about three quarters enclosed by the ciliated margin $c^1$. — *Original*.

Fig. 32. *a*, the same as fig. 31, whilst attached and fully expanded; *b*, a single individual with its posterior end partially coiled. 8 diam. — *Original*.

Fig. 33. The same as fig. 31, seen from the opposite side or back. 50 diam.

as to form a notch, and curls upon itself in a spiral form. The space ($s$) which is bounded by the rim ($c$) is called the *disc*. Close to the notch, and within the little spiral just spoken of, there is an opening (fig. 30 $m$, and fig. 31 $m$) which leads downwards into an elongated funnel-shaped channel (fig. 31 $m^1$); the first is the mouth and the second is the throat or œsophagus. The latter tapers to a point ($m^2$) after plunging for a short distance obliquely across and backwards into the centre of the body; and its further extent can only be detected when the food passes through it, on its way to the general digestive cavity.

Usually at about one third of the way around to the opposite side of the trumpet, but sometimes nearly half round, as in this figure, — by some peculiarity in the mode of expansion, — there is a globular cavity ($cv$) from which a tube ($cv^1$) extends, with a gentle taper ($cv^2$) to the extreme opposite end ($a$) of the body. By the rhythmical contraction and expansion of the cavity and the tube, you will recognize them to be the components of a contractile vesicle. The contraction of this heart-like body occurs once in three quarters of a minute, when the fluid in the globular portion ($cv$) is forced backward into the narrow channel, ($cv^1$, $cv^2$,) which, in its turn, after expanding very sensibly, and receiving the fluid, contracts, and returns it to the original starting-point. Looking upon that side of the trumpet at which the mouth is placed as the lower or abdominal side of the body, the contractile vesicle may be said to be normally situated at or about the left side.

The digestive and the circulatory systems are the only parts of the organization essential to life that are known to investigators; but recently I have been led to believe that I have discovered the *nervous system*, or at least a part of it, and that too in the very region of the body where there is the most activity, and therefore more likely than elsewhere to have this system most strongly developed. Immediately within the edge of the

The two resultants are just on the point of separating at $a^1$; $s$, the protruded convex outline of the disc; $m^1$, the throat in the distance; $cv$, the contractile vesicle next the eye; $n$, the reproductive organ. — *Original*.

disc there runs all around a narrow faint band, (fig. 30, $r^1$,) which lies so close to the surface that it is difficult to determine precisely that it is not actually superficial. From this band there arise, at nearly equal distances all round, about a dozen excessively faint thin stripes, ($r$, $r$,) which converge in a general direction toward the mouth ($m$). They are, like the circular band, just beneath the surface of the disc. With a microscope of ordinary quality, these stripes would be confounded with the longitudinal, closely set bands which run along the whole length of the trumpet. The circular band has been described by some authors as a circular vessel, which is connected with the contractile vesicle; yet, after the most careful examination, I have concluded that this view is not the true one, but rather as I have here described.

As I have said, the circular and converging bands of the nervous system are evidently in close relation with the most active organs. These organs are the mouth ($m$) and the vibratile cilia. The former I have described, and will therefore merely state that the convergent stripes appear to enter it and pass into the throat.

The vibratile cilia ($c$) are variously elongated, fine, short, hair-like bodies which cover the whole surface of the body. The largest, most prominent, and conspicuous of them are those which fringe the edge of the disc, and are arranged in a closely set series ($c$) along the outer margin of the circular nervous band ($r^1$), so that they appear to rise directly from it. There are also large rigid cilia ($b$) scattered over the body.

At certain periods there may be seen toward the right side of the stem a slender, ribbon-like, yellowish body ($n$, $n^1$), which extends lengthwise from near the disc almost to the basal end. Within this body, young ($n$) have been seen to originate, and therefore it is looked upon as the *organ of reproduction*.

With these facts in hand, we are now prepared to follow the process of budding, and the accompanying internal changes, with a clear understanding of their import. The first change that takes place, preparatory to this phenomenon, is a division of the contractile vesicle into two distinct organs. This separation takes place at about midway between the top and base of the body, (for instance, at $cv^1$, fig. 30,) and the upper end of the lower portion soon develops a globular cavity like the original one ($cv$) Soon after this there appears, directly to the right of the new globular vesicle, and exactly in the middle line of the ventral side of the body, a shallow pit, and around this pit a semicircle of vibratile cilia. The pit, which is the future mouth, gradually deepens, and a funnel-shaped cavity, the throat, is presently formed, which extends inwardly and backwards in the same relation to the body as does the œsophagus of the old head. At the same time, the semicircle of cilia increases in length, and takes a spiral course, in one turn about the new mouth, and then passes around to the left of the body, and obliquely forward until it reaches a point just behind the edge ($c$) of the disc, and midway between the ventral line and the left side.

This, I beg you will take note, is all a new growth, or budding, as it is called; and as yet no trace of external self-division is visible. But that follows now at once, and in about two hours, as I observed on one occasion, the separation is completed. For this purpose, the constriction commences considerably in front of the new mouth, and is most conspicuous, at the outset, on the left side. This seems, as it were, to push the new row of cilia further around toward the back, and at the same time obliquely toward the tail. When the constriction has passed across the body, from left to right, for about two thirds the way, the row of cilia has retreated, and extended around the body almost to its starting-point, so that it forms a broad open spiral, (fig. 31, $c^1$,) which has nearly the same obliquity as that around the old disc. By this time the ribbon-shaped reproductive organ ($n, n^1, n^2$), at the right of the mouth ($m$), has also divided. This is the condition in which the Stentor is represented in this figure, (fig. 31,) as it swam freely about in a partially closed state.

From this time onward, until separation, very little change takes place in the organization; the constricting simply narrows the yet remaining connection ($a^1$), and, as this is done, the two ends of the broad spiral ($c^1$) of cilia gradually approximate each other until they finally meet, (fig. 33,) just as the two individuals separate. These two resultants of the operation which has just been described, are to all intents and purposes perfectly identical in their organization, the only difference being that one is. in part, older than the other.

Now, as I have said before, there are all possible gradations between the methods of reproduction by budding and those by development from the egg ; and I intend next to illustrate the first step tending toward the latter process, although it is very far off from it. The resultants of the process, which is one of self-division, are unlike each other, and on this account the term *alternate generation* has been applied to this, along with other allied forms of reproduction.

I will introduce at once the two forms which represent the diverse extremes of the generation of one individual. You would not suspect by their figure (figs. 34, 35, and 36) that they were in the least related to each other; but yet the method by which the one is transformed into the other is perfectly simple. The Hydra-like individual, (figs. 31, 35,) with its short, trumpet-shaped body, and long, wavy, hair-like tentacles (*t*), was formerly believed to be a sort of marine Hydra, and received the name of Scyphostoma ; and although it is now well known to be one of the stages of growth of the Sunfish, yet it goes by the name of, or is spoken of, as the

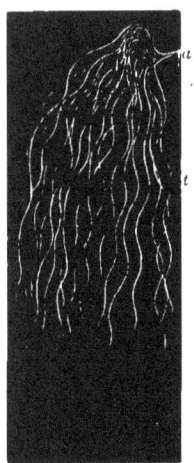

Fig. 34.

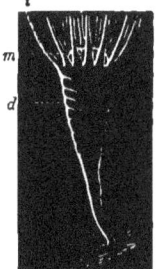

Fig. 35.

Fig. 34. The Scyphostoma of *Aurelia flavidula.* Per. and Les. Natural size. *a*, the base, attached to the side of the aquarium ; *t*, the tentacles hanging loosely, at their fullest extension, and waving to and fro with the movements of the water. — *Original.*

Fig. 35. The same as fig. 34, magnified eight diameters. *a*, the base ; *m*, the partially extended proboscis encompassed by the tentacles *t ; d*, the incipient formation of the Ephyras, foreshadowed by transverse rings. — *Original.*

Scyphostoma. In a general way, it is like the Hydra in structure, but when fully developed it has as many as twenty-four tenta-

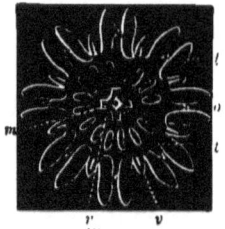

Fig. 36.

cles at the broad end of the body, disposed in a wreath at a short distance behind the mouth (*m*). From the laying of the egg to the full development of the Scyphostoma about eighteen months are required. Although but a few weeks are necessary for the development of all the tentacles, yet several scarcely noticeable features are much slower in appearing, and do not, even when completely established, produce a marked change in the external conformation of the Scyphostoma. Finally, however, toward the latter end of the second year, a very perceptible transformation takes place; the body increases considerably in size, and seems to be divided into several transverse rings, (fig. 35, $d_i$,) or superposed discs. This is brought about by an actual self-division of the body, which in process of time completely separates these discs from each other and from the original stock. Before this takes place, considerable change goes on in the discs; first, their edges become scalloped; then each of the little lobes of the scallops, growing longer, takes on a fork-shaped figure with two broad prongs, and, at the same time, broad channels or tubes are hollowed in the thickness of each disc, and a short funnel is developed from the centre, on the side toward the mouth of the parent. In this condition the discs, one after the other, break loose from the Scyphostoma, at its broader end, and swim freely in the water, each an independent individual.

At this period they go by the name of the Ephyra; a name which was originally given to them, before it was known that

Fig. 36. The Ephyra of *Aurelia flavidula*. Per. and Les. Six diameters. A view from the mouth, or oral side. *m*, the four-sided proboscis, with the mouth in the centre; *t*, the broad, marginal, digestive canal, connected with the central cavity (about *r*) by sixteen short, broad channels; *r*, the four incipient reproductive organs; *l*, the ocular lappets; *o*, the eye; *v*, the lancet-shaped veil, and the single, pointed tentacle.— *Original*.

they are one of the forms of the Sunfish, which had already received the name of Aurelia. The figure which I have here, (fig. 36,) represents this young Aurelia after it has been a free individual for a few days, and a slight change in its conformation has taken place. In shape it resembles a broadly spread umbrella, with a short, thick handle, the latter being the funnel which I mentioned just now. This funnel is a sort of proboscis which has the mouth (*m*) at the end of it. From the proboscis a passage opens into a broad, but not very deep central cavity, and from the latter sixteen wide tubes stretch out like the spokes of a wheel and join a broad circular channel (*t*), which passes around close to the edge of the umbrella. From the circular tube a short one passes into each of the eight, equidistant, double-lobed lappets (*l*). These cavities and canals constitute the digestive system, within which the particles of food circulate, backward and forward, through the agency of vibrating cilia, which are everywhere present, from the mouth to the ends of the canals in the lappets. Between every two of the latter there is a tongue-shaped body (*v*), and a single finger-shaped tentacle. The former is thin and pointed like a lancet, and possesses a great degree of extensibility and contractility, owing to its highly muscular development. It is the future marginal curtain or *veil* of the adult (fig. 37, *v*). The single tentacle is the first, and always remains the central one when the number is increased, in extreme age, to more than a hundred in each of the eight groups which alternate with the lappets (fig. 37, *l*).

At the bottom of the deep notch of each lappet there is a short, thick, finger-shaped body, (fig. 36, *o*,) with a bright, glistening tip. By the help of the microscope, we find that this glistening is caused by several closely set, minute, magnifying lenses, which to all appearances act on the light in the same way that the lenses in the eyes of insects do. On this account, these gem-tipped fingers are called the *eye-peduncles*, the group of lenses the *eye*, and the lappets (*l*) the *ocular lobes*.

The rudiments of the reproductive organs are four groups (*r*) of worm-shaped bodies which lie within the digestive cavity, at four points opposite, and close to, the four sides of the base of the proboscis ($m$).

It is not necessary for present purposes that I should go any further with you into the history of the development of the Ephyra; but I will, at any rate, indicate what is the final result of that process, in order that you may see to what extremes of difference the two kinds of individuals of Aurelia arrive.

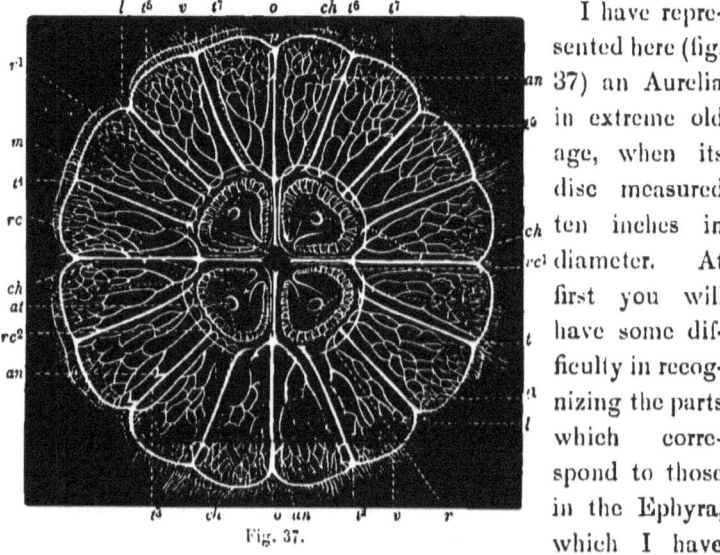

Fig. 37.

I have represented here (fig. 37) an Aurelia in extreme old age, when its disc measured ten inches in diameter. At first you will have some difficulty in recognizing the parts which correspond to those in the Ephyra, which I have just described; yet I think a word or two of explanation will make their relations clear to your eyes. In the first place I must

Fig. 37. *Aurelia flavidula.* Per. and Les. Natural size, 10 inches in diameter. The common "Sunfish" or "Sting-blubber" of the Atlantic coast, in its old age, when all of the digestive canals, both straight and branching, are united with each other. A view from the side opposite to the mouth, the whole internal organization appearing through the thickness of the gelatinous umbrella. The proboscis has four lance-shaped arms, here represented by the winding, dotted line, forming a cross, which extends to the edge of the umbrella, on the right and left and above and below. $m$, the central digestive cavity; $ch$, the four principal digestive canals; $t^1$, $t^3$, $t^5$, $t^5$, the four secondary digestive channels; $t$, $t^2$, $t^4$, $t^6$, &c., the eight digestive canals of the third set, at the points where they join the circular marginal canal, $t^7$; $an$, the points of junction, or anastomosis,

forewarn you that the proportions of the youngest Ephyra are not retained throughout life, but that they are constantly and gradually changing, from the fact that some organs grow faster than others. Starting from the central cavity ($m$), which is here seen through the thickness of the disc, there are, in the first place, four channels ($ch$) which pass in four different directions, opposite the corners of the proboscis, straight to the edge of the disc; secondly, there are four other less prominent openings, alternating with the outlets into the four channels just mentioned, which lead into these four very broad three-sided cavities, ($rc$, $rc^1$, $rc^2$,) that occupy nearly the whole area about the central digestive chamber. These last are especially devoted to the office of enclosing the reproductive organs,—which lie in the form of a deeply plaited ruffle ($r$, $r^1$) along the margin of each cavity,—and at the same time they allow the passage of the fluids from the centre toward the periphery of the disc, by means of several channels, of which there are, for each cavity, three principal ones, ($l$, $l^1$, $l^2$,) which go nearly direct to the marginal canal ($l^i$) of the disc, and three or four others, which form a net-work of tubes. These last are developed between the sixteen broad channels which appeared in the youngest Ephyra, and which are here represented by the same number of narrow straight tubes, ($ch$, $l$, $l^1$, $l^2$, &c.,) going alternately to the eight eyes ($o$) and to the middle ($l$, $l^2$, $l^4$, $l^6$) of the eight intervening groups of tentacles. Between the straight and the branching canals there is more or less of lateral communication, ($an$,) but not, however, until late in the life of the animal, at which period this figure was drawn.

The broad circular channel of the Ephyra is here a thin tube between the principal canals and the branching ones; $rc$, $rc^1$, $rc^2$, the four triangular cavities opening on the one side into the central cavity ($m$), and on the other side into the canals of the second, third, and fourth sets; $r\,r^1$, the frilled, ribbon-shaped reproductive organs lying, along the margin of the triangular cavities, in the distance, *i. e.* nearest the oral face; $at$, the opening leading to an excavation on the oral face and immediately opposite and nearly coextensive with each triangular cavity; $o$, the eye, bordered by the ocular lappets ($l$); $v$, the veil thrown outwards at the moment when the umbrella begins to contract. — *Original.*

($l'$) which follows the undulating margin of the disc, just within the rows of tentacles. The latter occupy eight broad spaces, between the eight eyes ($o$), and correspond to the eight narrow intervals between the ocular lappets (fig. 36, $l$) of the Ephyra. If you imagine now, that, as these intervals widen, new tentacles spring up on each side of the single one, and the lancet-shaped body ($v$) gradually broadens correspondingly, you will have before you, in time, the long rows of tactile organs and the pendent marginal veil (fig. 37, $v$) of the adult, stretching from eye to eye, throughout the circuit of the umbrella.

Comparing now these two individual forms, the hydra-form Scyphostoma (figs. 34, 35) on the one hand, and the medusa-form Ephyra (fig. 36) or the adult (fig. 37), with each other, you very naturally will inquire which one is to be looked upon as the original stock. The one which is called the adult produces the eggs from its reproductive organs (fig. 37, $r$); but then from each one of these eggs is hatched a Scyphostoma, which divides its single individual self into several disc-shaped, egg-bearing bodies, and, after throwing them off, retains its own integrity, to repeat the same act for an unlimited number of times. The question then is, are we to consider the Scyphostoma as the original individual, the grown-up single egg, which casts off periodically those parts of itself which contain the reproductive organs, to complete their further development, simply *as organs*; or must we look upon this hydra-form individual as essentially a composite being, a compound of several bodies, a part of whose number separate from the rest, and take on disguised forms?

If we had no other instances of self-division or budding on record, one could hardly refrain from adopting the latter view; but as it is, there are facts which lend a powerful argument for the adoption of the former category. Suppose, for instance, that an egg, instead of developing into a hydra-form body, expanded immediately into an ephyra-form, it would be a case in which but one medusa was produced from one egg; and we should then look upon the egg-stage as the one which corresponded to the hydra-form condition, and infer that the whole hydra became a medusa;* in other words, that the hydra was

* *The individuality of Hydro - Medusæ.* Perhaps the diversity of individuality is nowhere so extensively exemplified as among the Hydroida, nor tends so strongly to verify the belief in the merely partial individuality of that portion which is set free by budding or fissigemmation. It can be shown that the same individual stock bears evidence in its own person that what at one time may appear to be a distinct *individual*, at another time is incontestably the *reproductive organ.* As instances of this kind, I will adduce two of the most characteristic forms of reproduction that occur among Hydroids.

*Rhizogeton fusiformis.* Ag. The first that I shall describe is, in more than one respect, one of the

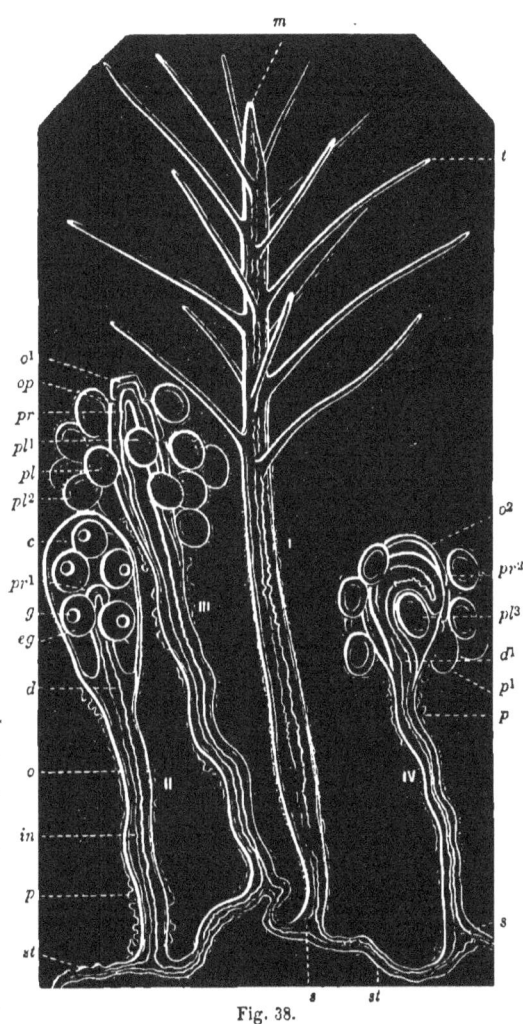

Fig. 38.

Fig. 38. *Rhizogeton fusiformis.* Ag. 20 diam. A marine Hydroid. I, A sterile hydra: II, III, IV, fertile hydras, or reproductive organs; *st*, the stolon, or creeping stem; *s*, point of junction of the upright stems with *st*; *m*, the mouth; $d$, $d^1$, the digestive cavity; *t*, tentacles; *p*, the parchmenty sheath; $p^1$, the

the medusa itself. Now it has been asserted that the eggs of a most deceptive, as regards its individualistic character, of all Hydroida. During the spring, there may be found, in the rocky tide-pools of our sea-coast, minute tufts of a tender, deep orange color, resembling some of the more delicate sea-weeds, at first sight. The little stems, of about one sixth to one quarter of an inch in height, appear to be diverse in kind among themselves to a manifold degree. The two most striking diversities are those of height; the more elevated stems (fig. 38, 1) bristle with slender needle-shaped bodies (t), and usually in point of numbers are one in every six of the two kinds. The shorter kinds are totally devoid of the bristling organs of the taller ones, and at the end vary from a smooth, club-shaped form (II) to a more or less globose shape (IV) with diverse degrees of knottiness (III and IV). All these various shapes and proportions are united in a colony of compound individuals by a creeping stem (st), which runs over the rocks, and forms the means of attachment for the whole group. The tallest forms might be compared to the workmen in a hive of bees, as their office is to obtain food for the whole colony, whilst the more diminutive sort, subserving the means of reproduction, would be the queen-bees and their immediate attendants. The feeders, or sterile hydras (1), consist of slender cylindrical shafts which taper at the free end (m), and likewise suddenly narrow appreciably where they join (s) the creeping stem (st). The mouth (m) is at the tip, and leads directly into a simple, narrow digestive cavity which extends to the base (s) of the stem, and thence in every direction throughout the creeping stolon (st) and into the fertile individuals (d, d¹). About one half of the sterile hydra, from its tip downward, is endowed with twelve to fifteen needle-shaped feelers (t) arranged in a spiral about the shaft. The body of this, as well as the creeping stolon, and the stems of the fertile kind, are composed of two walls (o, in), and the whole is covered by a thin, filmy, parchmenty, shrivelled envelope (p, p),

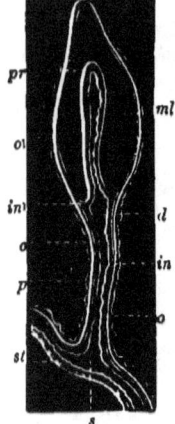

Fig. 39.

continuation of $p$ over the planules and the reproductive capsules; $o$, outer, and $in$, inner walls of the stems; $o^1$, $o^2$, the wall of the reproductive organs; $c$, the cavity of the same; $pr$, $pr^1$, $pr^2$, the proboscis-like body of the same; $op$, the fissures through which the planules ($pl$) escaped from the reproductive medusoid; $pl$, $pl^2$, planules on the outside of III, IV; $pl^1$, $pl^3$, planules within the same; $eg$, the eggs; $g$, the germinal vesicle of the egg. — *Original.*

Fig. 39. *Rhizogeton fusiformis.* Ag. A male reproductive capsule. $st, s, p, o, o^1, in, d, pr$, the same as in fig. 38; $in^1$, base of $pr$; $ml$, cavity of the capsule. — *Original.*

certain medusa, a Pelagia, develop precisely in this way.* If, then, by some process, that part of the Ephyra of Pelagia in which may be traced even to the tips of the feelers ($t$), and very easily followed all over the club-shaped ends ($p^1$) of the fertile individuals.

In order to understand the true character of the latter forms, we must study the structure of the smooth, club-shaped individuals (II). This clubbed termination of the stem is simply an expansion of the latter in such a way that the outer wall ($o$) separates from the inner one ($in$), and leaves a space ($c$) between the two. Within this space, and consequently around the inner wall which projects into it like a proboscis ($pr^1$), the *eggs* ($eg$) are developed. This is plainly then a *reproductive organ* raised on a stem identical in character with that which supports the feeding hydra (I).

Did the colony consist merely of a single feeding hydra and a group of fertile ones attached to it, — for instance, just as it appears to be in this illustration, — it would be no difficult matter to decide that the feeder is the individual centre, and that the fertile ones are its reproductive organs, which are attached to it by elongated pedicels, in the same way that the reproductive organs, or medusogenitalia, of Tubularia (chapter XIV.) are joined to the head. The relationship of the organs in question is the same in Tubularia as in Rhizogeton, with only a difference in degree, which is bridged over by the intermediate forms of Coryne (fig. 42) and Clava, in which the meduso-genitalia are more or less scattered along the shaft of the hydra, which represents in itself both the feeder and the reproducer. But the character of the individuality of Rhizogeton is complicated, inasmuch as it is a compound of several feeders and numerous reproducers, so that one could not tell to which of the former any one of the latter belonged.

As I said before, the nature of some of the fertile forms cannot be understood without a close study of the smooth, club-shaped (fig. 38, II) ones. The relations of the walls of the latter I have already pointed out, and it now remains to

---

* There happen to be two well-authenticated instances on record: one in regard to the Steganophthalmic, and the other respecting the Gymnophthalmic Medusæ. In both cases the discoverers traced the development of the medusa *directly from the egg*, and thus put the facts beyond a doubt. See Krohn, " Ueber die frühesten Entwickelungsstufen der *Pelagia noctiluca*," Müller's Archiv, 1855, p. 491, Taf. XIX., and Claparéde, " Ueber geschlechtliche Zeugung von Quallen durch Quallen," Zeitsch. für. wissenschaft, Zoöl., 1860, vol. x. p. 401, Taf. XXXII. figs. 1, 2, 3. The observations of Gegenbaur, J. Müller, Fritz Müller, and L. Agassiz, upon the development of medusæ without an intermediate hydraphase, are open to doubt, inasmuch as the young were not traced from the egg; and therefore no one can say that they did not arise by budding, either from a hydra-form, or from the internal surface of the digestive cavity of a medusa.

which the reproductive organs are situated were to be separated from the body and left to take care of itself, we would have pre-

trace them in the knotted forms. In course of time the eggs ($eg$) are metamorphosed into the secondary or planula stage, and the young in this condition, resembling opaque orange-colored globules, escape from the sac in which they originated. At the time, in May, when I discovered this peculiar form of reproduction, the egg-sacs (II), of a club-shaped figure, were comparatively much smaller and far less conspicuous than the knotted form (III, IV), and consequently the latter attracted my attention first. Most prominent among these were those represented in No. III of this group, and at first they were mistaken for the feeders which had lost their feelers, and whose meduso-genitals were crowded all over the head. Presently, however, a more expanded form (IV) was met with, which had one ($pl^3$) of the young still within the space between the outer wall ($o^2$) and the proboscis ($pr^2$), and the latter was partially coiled around it. This revealed the character of the knotted heads, for it had also on its outside the globular planulæ ($p^1$) which gave it the irregular outline that first attracted my attention. Returning then to the tallest and most slender of them (III), I found some in which, although the outer ($o^1$) and inner ($pr$) walls were nearly as closely approximated as in the sterile hydra, there was yet one young planula left ($pl^1$); and in addition, to clear up the mystery, the fissures ($op$) in the outer wall, through which the young escaped, were quite conspicuous at two or three points.

Why, then, were the slender forms mistaken for feeders which had lost their feelers, and why did they seem to be covered with meduso-genitalia? Simply because that when the planules escape through the fissures ($op$) of the meduso-genital, they are kept from dropping away from it by the thin, filmy envelope ($p, p^1$), which, it appears, has a great degree of ductility; and very naturally, from their position, the planules appeared to be groups of closely set reproductive sacs, *i. e.* meduso-genitalia, with exceedingly short stems for attachment to what appeared to be the head of the hydra. The whole process of this affair may then be summed up thus: the eggs maturing and changing into planules, the latter escape in succession through the outer wall of the sac; and as this goes on, the sac shrinks and approximates the inner wall (III, $pr$), until, when all the planules are expelled, the two walls ($o^1, pr$) lie close together, apparently as in the sterile hydra, and thus this meduso-genital appears to have all the characters of the latter, excepting its feelers.

What is the subsequent fate of the fertile form, I have not been so fortunate as to ascertain; nor has any observer taken note of it; although it has been assumed, and in fact positively asserted, that it finally becomes " metamorphosed into a hydra." (See in Agassiz's " Contributions," vol. IV. p. 226, the editor's totally unwarrantable assertion, *intercalated in italics* between my sentences, in

cisely what occurs when the Ephyras of Aurelia separate, and bear off the reproductive organs from the hydra-form. Yet, for

the description of the male of this Hydroid. Fortunately for science, the sentence beginning four lines below this was overlooked, and remains *unaltered* as a crushing evidence against the wholesale assumption just above it. * * * " *the specimens died* " * * * !!) I have given it as my opinion elsewhere, (Proceedings Boston Society of Natural History, Dec. 2d, 1863,) that "the probable explanation is that the medusoid" (meduso-genital) "withers, and becomes resorbed, and then a hydroid head develops directly from the end of the old stem."

*Coryne mirabilis.* Ag. The other Hydroid which I propose to describe here is known as *Coryne*, and is much more common than the first one, not only along our coast, but even in the brackish waters at the mouths of rivers, attached to rocks and floating timbers, &c. It varies in height and breadth according to the age of the colony, but usually the whole branching group is not more than half an inch high (fig. 40). The upright stems ramify more or less, and the end of each branch bears a club-shaped head (fig. 42, *m* to *t*), which is covered by a complicated spiral row of globe-tipped feelers (*t*). The *reproductive organs* (*pr*, *pr*¹) are attached to the head behind, or occasionally among the group of feelers. During the earlier part of the breeding season these organs develop into highly organized bodies (fig. 40, *b*), and finally break loose from the head of the hydra and swim away in this form (fig. 40, *a*), when it is known as the *medusa*. It seems to have, for all intents and purposes, a perfect organization of its own, excepting that it exhibits no trace of a reproductive organ. When just set free it measures about one sixteenth of an inch in diameter (fig. 40, *a*), but in process of time it grows to a diameter of half an inch (fig. 41), and at that size it has fully de-

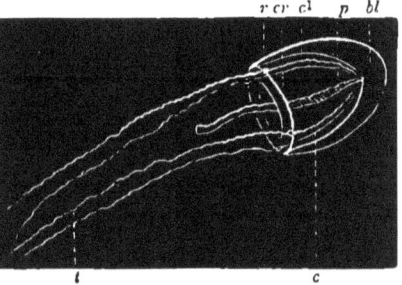

Fig. 40.

Fig. 41.

Fig. 40. *Coryne mirabilis.* Ag. Natural size. A group of *hydro-medusæ*, at the height of the breeding season; *b*, the meduso-genitalia; *a*, the same just set free, in the act of swimming. — *Original*.

Fig. 41. *Coryne mirabilis.* Ag. A full-grown *meduso-genital*, natural size, in the act of swimming with the tentacles (*t*) trailing in the wake; *bl*, the gelatinous bell; *p*, the proboscis projecting through the aperture of the bell; *c*, *c*¹, the four longitudinal canals; *cr*, the circular canal; *r*, junction of *c*, *c*¹, and *cr*. — *Original*.

all this, we must look upon what is left and what is taken away as two peculiar kinds of individuals, inasmuch as each leads an independent, self-sustaining life.

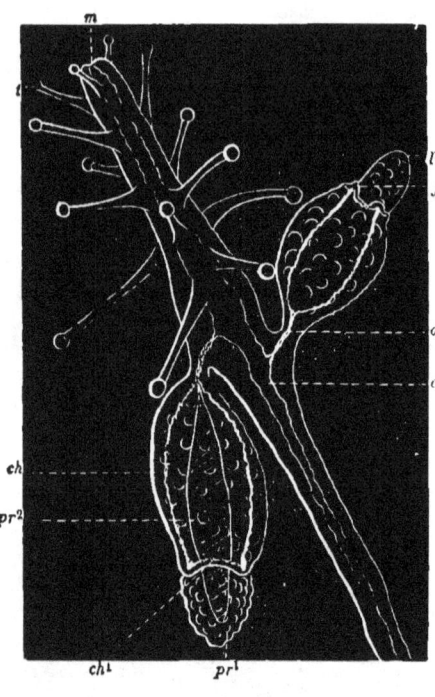

veloped eggs imbedded in the substance of the slender body ($p$) which hangs down like the tongue of a bell. The main portion of it consists of a jelly-like, bell-shaped body, ($b$) within whose thickness four canals ($c$, $c^1$) are hollowed along four equidistant lines, from where the tongue (proboscis $p$) of the bell is attached, to the edge, where they join another canal ($cr$) which goes around it. At each point of juncture ($r$) of the four canals and the circular one, a long, slender hollow feeler ($t$) is attached. The proboscis ($p$) takes food in through an aperture, the mouth, at its end; the food is digested within the same, and then, in a fluid state, circulates through the canals and into the hollow feelers.

Fig. 42.

Putting together now the organization of this body and the digestion of food which is carried on within it, there would not seem, at first, a possibility for a doubt even that the medusa is a perfectly independent organism, an *individual;* but a few days' acquaintance with the career of these budding bells, during the latter part of the breeding season, brings a totally different interpretation to one's mind. Along in March, (see Agassiz's " Contributions," vol. IV. p. 189,) I

Fig. 42. *Coryne mirabilis.* Ag. The head of one of the hydroids, late in the breeding season, magnified 50 diameters. $m$, the mouth; $d$, digestive cavity; $d^1$, prolongation of $d$ into the reproductive organs; $t$, tentacles of the hydroid; $pr$, $pr^1$, the proboscis-like body of the reproductive organs crowded with eggs; $pr^2$, the hollow of $pr$, $pr^1$, communicating with $d$ through $d^1$; $ch$, the longitudinal, and $ch^1$ the circular canals corresponding to $c$, $c^1$, $cr$, of fig. 41; $f$, the rudimentary feelers. — *Original.*

In the next illustration of self-division the two resultants seem to be more truly individual in character than those which we have just been talking about; but a similar phenomenon in

had noticed that the medusæ no longer broke loose from the head, and that at the same time they did not develop quite so fully as earlier in the season. A little later I found that they were still more undeveloped; the feelers were mere conical projections, and the bell was much deeper, like a thimble in shape; but what most particularly struck me was that the proboscis was much larger than in the free medusæ. In the course of a few days I obtained an abundance of Coryne upon whose heads all the medusa-like bells possessed an enormous proboscis (fig. 42, $pr$, $pr^1$, $pr^2$); in fact, large enough to fill the whole cavity of the bell and project far beyond it. The greater bulk of the proboscis consisted of closely crowded eggs, imbedded beneath its wall. The feelers ($f$) were scarcely if at all developed, and the canals ($ch$, $ch^1$) were in a more or less evanescent state. In fact the whole organization of the bell corresponded in every essential respect with similar organs of another Hydroid, (Tubularia, chapter XIV.,) which never possesses them in a more fully developed state, but loses them by a process of withering away, after the eggs have developed into moving bodies similar to itself, and escaped.

No one would hesitate to call the egg-producing bell of Tubularia a reproductive *organ*, nor would it seem illogical, therefore, to insist that the identical organ of Coryne should be looked upon in the same light. That it develops, at an earlier season, to a higher degree, and becomes a freely moving body, does not make it any the less a reproductive organ; for if it did, where then are these organs in the mean while? Acknowledging that when in its free state it has an individualistic character, yet nevertheless it is the *reproductive organ*, and *the only one*, as truly so as the always persistent, medusa-like, egg-bearing organ of Tubularia.

The reproductive organ of Rhizogeton (fig. 38) is identical in character with that of Coryne, but its development never carries it beyond that of a simple sac, without canals or feelers. Now between the degree of development of the reproductive organ of Rhizogeton and that of the highly complicated Coryne, there are all possible shades of gradation, many of which I have shown to occur even in the latter animal, and, as a consequence, there are likewise as many degrees of individuality, *if* it is insisted that these egg-producing bodies are not parts of the hydra organism, but essentially and actually separate individual medusæ. This conclusion is inevitable, for it cannot stand to reason that on the one hand the mere fact of breaking loose from the hydra constitutes the individuality, whilst on the other hand those meduso-genitals which, although developing to a scarcely less degree, remain attached, are consequently rendered mere appendages, or *parts of the organism* of the hydra.

closely allied animals presents one of the resultants in such a form, that, were it the product of the only known instance of self-division, it would be set down, without hesitation, as simply the reproductive organ removed from the body of a perfect individual. The latter is the Tape-worm, of which there are a great number of species living in, and undergoing this process in the intestines of all sorts of animals, and the former is one of the marine worms, of which, as well as of the fresh-water kinds, many have been found to undergo self-division.

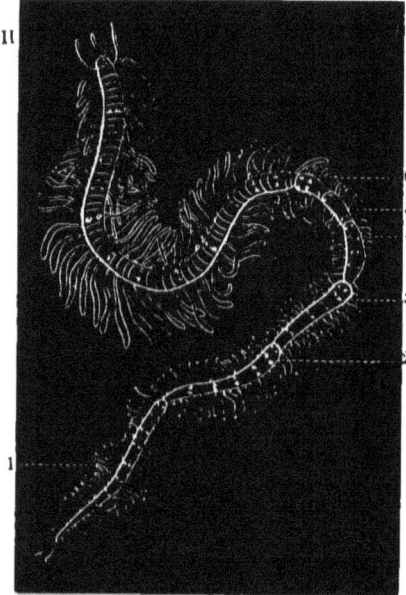

Fig. 43.

The most remarkable example known among the aquatic worms is represented here (fig. 43), under the name of *Myrianida fasciata*, as the species was originally called by its discoverer.* The oar-like appendages which are attached to the rings of the body are devoted not only to the office of propulsion, but also, receiving branches from the blood-vessels of the body, serve the function of gills, or breathing organs. The mouth is at the broader end, and from that the intestine extends in a direct course to the extreme posterior end of the body. About the intestine there are several blood-vessels which cover it with a sort of net-work, and, as mentioned just now, are connected with each of the branch-

Fig. 43. *Myrianida fasciata.* M. Edw. 2 diam. A marine worm, undergoing multiplication by transverse self-division. II, the head; 1, 2, 3, 4, 5, 6, the successively more advanced individuals, 1 being the oldest and 6 the youngest. — *From Milne Edwards.*

* Milne Edwards, Voyage en Sicile, vol. I. p. 43, pl. VII. fig. 65.

ing vessels of the gills. Usually there is, in worms of this kind, one vessel which is larger than the others, and which runs above them, in a straight line, to a greater or less extent, along the back of the intestine. This is usually called the heart. On the lower side of the body, and along its middle line, there runs a nervous cord, which is swollen into knots or ganglions, as they are called, in each joint. The principal ganglion (the brain) of this organ, is situated in the head (*II*) above the intestine, and is joined to the ventral cord by a sort of collar, the *nervous collar* so called, which embraces the throat. The reproductive organs are to be met with only in the posterior rings of the body, and are located there apparently in reference to the peculiar mode of reproduction which I am about to describe.

Nearly a century ago a Danish naturalist, Müller, had observed that a certain marine worm divided itself into two parts, and in some instances into four, and that each part continued to live after the separation. He moreover noticed that the anterior or oldest part of the worm did not contain any eggs, but that the latter were altogether confined to those parts which separated from it. Later observers, in several instances, have confirmed the statements of Müller, and, among others, Milne Edwards has added largely to our knowledge of this phenomenon by his elaborate investigation of the sixfold fissigemmation* of Myrianida. He noticed, what Müller had already mentioned, that, where more than one young individual was formed, the oldest was always the last (1); but what made the case the more striking in Myrianida, was that there were no less than six young developing at one time, and that they were, in regular succession (6, 5, 4, 3, 2, 1), older as they were farther from the head of the original stock. Each was possessed of a pair of eyes at the anterior end, and of a certain number of rings or joints. Thus the youngest (6) had ten joints; the second (5) fourteen; the

* A word compounded of *fissus*, divided, and *gemmatus*, budded; in allusion to the peculiar process here involved, which consists of a growth by *budding*, and a multiplication of individuals by the *division* of the compound body.

third (4) sixteen; the fourth (3) eighteen; the fifth (2) twenty-three; and the sixth (1) and last one, thirty. The process of the formation of these successive individuals is described as a sort of intercalation of joints; thus the joints of the first-formed (1) young are developed between the last joint of the old stock and the ring before it; then the next oldest, or second young (2), is formed by the successive appearance of a series of joints between the head of the first young and what was the next to the last joint of the parent; the third originates in the same way between the second young and the parent, and so the fourth, fifth, and sixth; and as each of the successively formed individuals continues to add new rings to its own body, the oldest has of course the greatest number, and very naturally is the first one to assume an independent life, whilst the others follow, according to their age, until all are separated.

We have thus a natural cutting apart of a set of organs, such as the intestine, the blood-vessels, and the nervous system, which, in some respects, are of a more highly developed character than those of Insects; and yet there was a time when it was thought to be nothing short of a miracle that even such a simple sac-like animal as Hydra could multiply itself by a partition of its single body. We would seem to be impelled here, too, to look upon these several resultants of Myrianida as more truly individualistic than those of the Scyphostoma (p. 67), because the organization of the former is apparently of a more highly developed and complete character than that of the latter. In both cases the newly developed individuals are the egg-bearing ones, whereas the original stock is sterile.

In the *Tape-worms* (fig. 44), the egg-bearing joints (*a* to *g*) separate from the old stock by hundreds, but each joint acts as an individual, and enjoys an independent existence; yet of such a questionable character, however, that one would almost seem to sacrifice his common sense were he to assert that these single joints are each a self-sustaining, self-conscious being. The most that we can say is, that while the fissigemmate progeny of Myrianida approach nearest to the idea of a fully developed, complete individual, the joints of Tænia stand the farthest off.

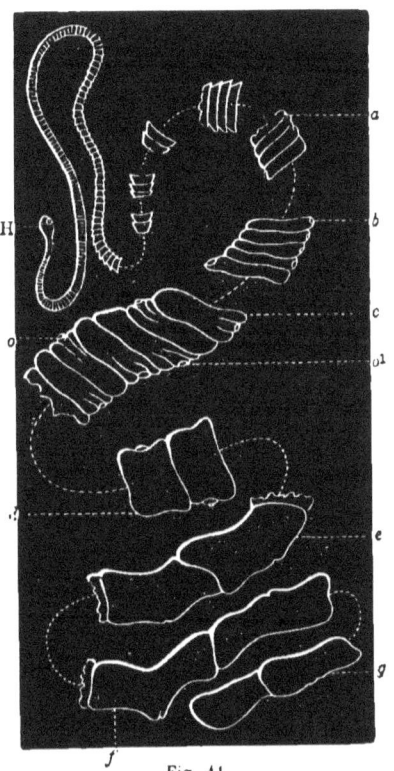

Fig. 44.

That you may the more fully understand this, let me state that Tænia has no intestine whatever, and that it possesses the merest trace of a nervous system, only a simple knot or ganglion which gives off a slender branch to each one of the four suckers that project from the four corners of the head (H). The only organ which seems to be common to the whole body is a sort of circulatory system, consisting of four slender channels, which, beginning at the head, extend, one on each side above and below, along the entire length of the animal, and communicate with each other from point to point by a transverse canal, which runs across the forward end of each joint. In some kinds of tape-worms these channels are distributed in the form of an irregular net-work, which here and there has minute openings at the surface of the body, by which they communicate with the surrounding fluids of the an-

Fig. 44. "*Tænia solium*. Linne. Natural size. Only such parts are represented as are characteristic for the shape of the joints. H, the head; *a*, 309th joint; *b*, 448th joint; *c*, 569th joint; *d*, 680th joint; *e*, 708th joint; *f*, 849th joint; *g*, 855th joint, and last but one. This worm measured 3200 millim. (10 feet 9 inches)." *o, o¹*, the apertures of the reproductive organs. — *From Weinland*.

imal in which the worm lives as a parasite. Whether this is a circulatory system, or one which serves to bring the surrounding medium into more intimate contact with the internal tissues, in order that it may be absorbed as a nutriment, is still a question unsettled among naturalists. Whatever food is taken by the Tænia is introduced, no doubt, by some method analogous to that by which a sponge absorbs water. When, therefore, the joints are set free, there is merely a cutting in two of one of the few organs which the body possesses; and each of these individual segments consists, then, of a reproductive organ * and a part of the system of the canals of the parent. The latter do not serve any purpose whatever, since the joints soon cease to show signs of life, and, in fact, decay not long after they are discharged from the animal in which they lived, and have expelled the eggs from the ovary. They are, functionally, *egg-sacs*.

With this knowledge, then, of the structure of the joints of Tænia, we are justified in looking upon them as enjoying the remotest degree of individuality, when contrasted with that of Myrianida; and were there not instances, among the different kinds of intestinal worms, of all possible grades between these two extremes, we should not hesitate to throw that of Tænia altogether out of the category of individuality, and set it down as one of the many known modes of laying eggs.

Putting aside, now, these questionable forms of individuality, which, by the way, I would say, I have discussed at pretty full length for the sake of future reference, we have in such animals as Hydra, (fig. 27,) Metridium, (fig. 28,) and Stentor, (fig. 30,) examples of the formation of perfect individuals, without the intervention of the egg-phase. I will state, moreover, that these are by no means single instances, for there are several species of Hydras, and the kinds of Anemones are numerous all over the globe, whilst of the group of Protozoa, to which Stentor belongs, there are hundreds of species in various lands, many of which we know, from actual observation, do undergo this

---

* Each joint is really compounded of twins, one a male and the other a female.

process; and of the others, from their close affinity, we cannot doubt that they reproduce themselves in the same way. This is, in one sense, then, a negation of the old aphorism: "*Omne vivum ex ovo.*" "*All life comes from an egg.*"

I have not, by any means, begun to exhaust the numerous instances in which individuals originate without passing through the condition of an egg. The process of budding prevails not only among the lowest forms, as I have shown, but it occurs among animals which are higher in grade than the Worms; it occurs among certain shrimp-like animals, called "water-fleas," which are related to crabs and lobsters; and it occurs among Insects. It is common among certain forms of Shell-fish, Bryozoa, (see chap. xi.) Finally, we have it even in Vertebrates, especially among Fishes; as Lereboullet has recently shown to prevail to a great extent. Sometimes, according to this observer, an egg divided into two fishes; sometimes the division was only partially carried out, and the result was a fish with two heads, (fig. 45,) or two tails, or two heads and two tails and the body single. It is very seldom that full-grown fishes are found in this condition, because they are more or less helpless and unable to defend themselves or escape from an enemy. It is not a very rare thing to find full-grown snakes with two heads or two tails. Dogs, cats, calves, &c., have been born, and have grown up, with an increased number of heads or legs; and even Man seems not always to have been contented with one head. Isidore St. Hilaire has recently published a work in which he has given a great number of examples

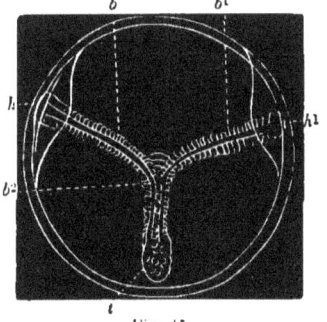

Fig. 45.

Fig. 45. *Esox lucius.* L. A double-headed embryo of the Pike of Europe, magnified 20 diam. *h, h¹*, the two heads; *b, b¹*, the anterior half of the body of each; *b²*, the junction of *b, b¹*, which continues single to the tail (*t*). The whole lies on the surface of the yolk, and is as yet enclosed within the egg-shell. — From Lereboullet.

of this kind of human duplicity; and I might add, also, not only a duplicity of the head, but also of the heart, and of the tongue. The Siamese twins are, probably, an example, where the egg did not altogether separate into two parts. No doubt the many instances of twins, and triplets, and quadruples, which occur so often lately, are partly derived from this method.

This phenomenon, however, occurs most frequently, and, as a *rule*, among the *lower* animals; and as we ascend in the scale we find that it happens less often, and rather as an *exceptional* mode, and finally, its appearance among the highest animals is rare, and is looked upon as a *monstrosity*. I think, however, that this term, *monstrosity*, has served to mislead physiologists; and so great oftentimes is the power of a name or term, — which may be persistently thrust forward by some one who has a speciality or lifelong hobby to support, — that it has prevented the younger class of observers from attempting to look through this thick curtain into the realities behind it.

Now it is a noticeable fact that these duplications and buddings which occur among the *highest* animals make their appearance by far most frequently among the lowest and diseased classes, where the qualities of life are in the most degraded condition. It would seem as if the individuality, the *oneness*, of the healthy, *natural* man, loses its power of concentration and tendency toward a higher life when disease ensues, and especially when it is propagated from parent to child and child to grandchild; and that Nature then reverted more or less toward the lower, more degraded kinds of existence, and very naturally reproduced herself according to a method which corresponds to this reduced condition, in which there is an approximation to the state of vitality and grade of the lower creatures on this earth. This, I am inclined to believe, is the true explanation of the phenomenon which is called monstrosity among the highest kinds of animals.

But to return now to the more simple phases of nature. I

repeat it, there are individuals which habitually, and as a *rule*, originate without the intervention of the *egg-state*. And now why may there not be yet another undiscovered method by which animals originate, more simple in character than those three already known?

## CHAPTER IV.

THE REGENERATION OF LIVING ORGANISMS AFTER PARTIAL DESTRUCTION. — THE PERSISTENCY OF VITALITY DURING DECOMPOSITION. — THE RELATION OF SECONDARY CAUSES TO THE GREAT PRIMARY CAUSE. — ANIMALS PRIMARILY CREATED IN AN ADULT STATE.

Having demonstrated to you that the egg is merely one of the early states of existence, and not a body which preëxists the animal proper; and having shown that there are certain creatures which, even in their adult state, are as simple in structure as the least complicated eggs; and finally, that the so-called egg is left more or less dependent upon physical causes for its growth, in some cases nearly altogether independent of the influence of the parent, and consequently dependent upon physical agency for its growth almost entirely; I remarked, then, that there seems to be but one step left to bring us to that condition of things under which the *animal-egg* may arise altogether independent of a parent. I then went on to state that it is an acknowledged fact among naturalists, that some individuals do not originate from the *egg*, but by a process in which both male as well as female reproduce their like, that is, by budding or by self-division; and exemplified these processes by the budding of Hydra and the Anemone, and by the fissigemmation of Stentor, the Hydra-like young of Aurelia, Myrianida, Tœnia, and some of the higher forms of creation, such as the fishes and reptiles, and the warm-blooded animals, Man not being excepted.

Now it having been shown that there are three different modes by which an individual existence arises, the question further demands whether there may not be still another yet undiscovered method, even more simple than those already known.

Furthermore, these buddings of Hydra and Metridium are mere subdivisions of cells; one cell alone may subdivide; an egg which is a single cell may do the same; and so may an animal, which is essentially a single cell, do likewise. By budding, we see that animals do, in a measure, arise independent of a parent; certainly independent of the egg-stage; and therefore, as all individuals do not originate by maternal gestation, we cannot be debarred from inquiring how many other modes of generating animals there are. We have already taken note of one, namely, that by budding; and of another, namely, that by self-division. Let us see, now, to what extent reproduction by *artificial division* proceeds; under what forms, and in what or how many ways and conditions it happens.

We have seen how small a piece of the base of an Anemone (p. 57) may divide off voluntarily; now I will add that it may be cut across and the base reproduce a head, and the head reproduce a base; or it may be split lengthwise, and each half will regenerate the wanting part. Infusoria were cut in several pieces by Ehrenberg, and each fragment reproduced what was wanting to complete its organism.

*Trembley's experiments upon Hydra.* — The most remarkable of all these kinds of experiments are those of Trembley, upon the Hydra, of which he published an account in 1744. Had they not been confirmed by other observers and experimenters, there is no doubt that the statements of Trembley would have remained in obscurity, along with the stories of the old writers about the now justly termed fabulous monsters, the Griffins, the Serpents with many heads, which were called Hydras, the Tritons, Centaurs, &c. Not only did this patient experimenter cut the Hydras in two, but he even went so far as to slice them across into numerous thin rings, and, marvellous to say, even at this day, each ring reproduced a crown of tentacles at one end, and elongated into a perfectly formed, naturally shaped individual. With the same degree of minuteness, Trembley also split the Hydras into thin longitudinal strips, which, like the rings, reproduced what was wanting to make a perfect body. Some of them he split from the mouth only part-way down the body and, each part reproducing what was needed, a many-headed Hydra was the result; thus verifying, on a small scale, the story of the many-headed monster of olden time. Yet the ingenuity of Trembley was by no means exhausted; for seeing that these little creatures were mere sacs, the idea of turning them inside out struck him as a feasible one, and he therefore proceeded to the experiment with a great deal of care and perseverance, as his detailed account of his method of performing the act seems to show. With the help of the blunt end of a fine needle he pushed the bottom of the sac through the body and out of the mouth; but he found that the animal righted itself as soon as left alone again, and therefore, after having again succeeded in the inversion, he ran a bristle crosswise through the body, and thus compelled the little victim to retain its change of front, and reorganize its internal and external departments. This it did not seem to have any difficulty in accomplishing, after the lapse of a few days, as Trembley proved by presenting it with bits of meat, which it swallowed with accustomed voracity.

Trembley now undertook to ingraft one individual upon another; and this he succeeded in doing after some curious experiences. At first he pushed the tail of one individual deep down into the cavity of another, and in order to hold them in this position he ran a bristle through their bodies, (fig. 46, A,) and tied a knot in the end which was below the surface of the water to prevent the spitted pair from leaving their post.

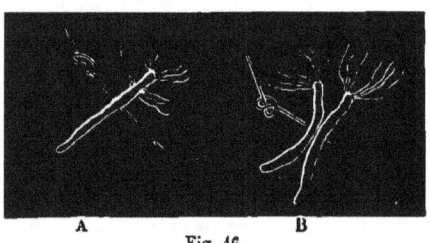

Fig. 46.

But the simple Hydras outwitted their tyrant, who, to his great amazement, found them, some hours after, hanging side by side, (fig. 46, B,) as if they had never been under more intimate relations. He concluded, then, to watch the next pair, when he discovered that the inner one first pushed its tail through the hole

Fig. 46. *Hydra fusca.* Trembley. Twice natural size. A, two hydras ensheathed one by the other, and a bristle run through them. B, the same as A, separated from each other, but still spitted by the bristle. — *From Trembley.*

made by the bristle, and then drew its head after it, and sliding sidewise along the spit, completely freed itself from its companion. This it did as often as the experiment was tried in that way. It then occurred to Trembley that dissimilar surfaces, that is, the outside of the one and the inside of the other, were not so likely to grow together as similar ones; and to put this to the test, he turned one of the Hydras inside out, so that when it was pushed into the cavity of the other, the surfaces of the stomachs of both were brought into contact. With this condition the animals did not seem to be dissatisfied, since they remained as they were fixed, and finally united themselves in one body, and enjoyed their food in common.

Not only, therefore, do we see that these lowly organized creatures may reproduce a lost part, but that they are able even to resign their individuality, in order to fit themselves for the conditions in which they are placed. It would seem from this, moreover, that the rank of individuality in these beings is of a very low order, that they should so easily part with it, or merge it into one of a complex and indefinite nature; and it would consequently appear to confirm what I have already said in regard to the insensible gradations from the lowest to the highest degrees of individuality.

But to proceed with the enumeration of other cases of artificial division, I will mention one which you can very easily verify in the common starfish (chap. x. fig. 109) of our coast. It often happens that a specimen is found having one of the arms much smaller than the others, and now and then with two or three in this condition; and rarely with only one arm out of the five left, the others being represented by little points growing out of the old stumps. Some kinds of starfishes suffer so little violence from the breaking of their arms, that, let them be ever so gently handled by their captor, every limb will drop off voluntarily before the specimen can be transferred to an aquarium. Of our common starfish I have, on different occasions, kept specimens alive after breaking off all but one of the arms, and seen the wounds heal over. I was not so fortunate, however, as to preserve them in a healthy condition in their confined state, and therefore did not actually watch the daily growth of the new arms.

In an experiment upon a more highly organized animal, I was better favored, and succeeded, after cutting it in two across the middle of the body, in obtaining a most ample proof of its regenerative powers. It belongs to the class of Worms, and is

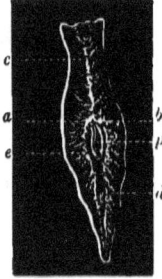
Fig. 47.

known as the *Flat-worm*, or Planaria, (fig. 47). There are numerous species in our ponds, where they creep over the surfaces of stones and aquatic plants. The one which is figured here is very common, and readily detected and recognized by its opaque white color, and the liver-colored ramifications of its intestine. The mouth is at the middle of the body, on the under side, and opens into a short, thick, cylindrical proboscis, ($p$,) which when retracted lies in an oval cavity. From the latter, the intestine extends in two, or rather in three directions; anteriorly ($c$) it reaches, in a median course, nearly to the end of the head, and gives off numerous, irregular, lateral branches toward the sides of the body; in the opposite direction it proceeds in two nearly parallel lines ($e$) backwards, on each side of the proboscis, until it thins out into very minute branchlets in the tail. Each limb gives off numerous ramifications toward the edge of the body, on that side in which it lies. The reproductive organs are situated mostly in the posterior half of the body, and their opening ($d$) lies half way between the proboscis and the end of the tail. The nervous system consists principally of an oval mass, which lies across the anterior end of the body, and two slender, thread-like, indistinct nerves, which extend backward along the lower side of the intestine. If, now, the animal is cut in two, at a point (through $a$, $b$) just

Fig. 47. *Dendrocælum lacteum.* (Œst.?) 3 diam. A milk-white, fresh-water Planaria. $p$, the proboscis; $c$, the anterior branch of the intestine; $e$, the posterior branches of the intestine; $a$, $b$, point of junction of $c$ and $e$; $d$, the aperture of the reproductive organs. — *Original.*

behind where the two posterior branches of the intestine part from the proboscis, ($p_1$) the anterior half of the body will, in order to become perfect again, have to reproduce the posterior branches of the intestine, a new mouth and proboscis, and the whole of the reproductive organs; whilst the posterior half must do the same for the anterior branch of the intestine, ($c_1$) and the main part, the so-called brain, of the nervous system. These two figures (fig. 48, A, B) represent the condition of the two halves after they had been cut apart eleven days. On the 24th of September the division was made with a sharp knife; and whilst the animal was under a magnifying lens, in order that the precise line of the cut might be seen clearly. The process of reproduction was watched from day to day, and the method of procedure carefully noted. As has already been stated by others who have performed this experiment, the two halves crawled off, after the section, as if nothing had happened; the anterior part preceding an imaginary tail, and the posterior one following an equally ideal head and brain. By frequent observations upon the newly developing parts, I ascertained that the restored organs were not formed all at once, but gradually, as it were bit by bit, in this wise. From the anterior half (A) a point insensibly budded out at the cut end, and within this projection a clear spot ($m$) appeared, which eventually proved to be the retiring chamber, or sheath of the proboscis when retracted within the body; next the proboscis, ($p_1$) with a gradually defining outline, made itself apparent, and at the same time irregular

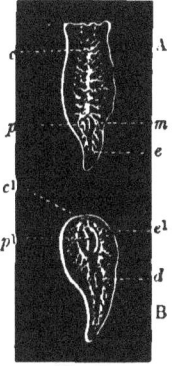

Fig. 48.

Fig. 48. The same as figure 47, after being cut in two and allowed to regenerate for eleven days. A, the anterior half; $p$, the new proboscis; $m$, the cavity into which the proboscis retracts; $c$, the original anterior branch of the intestine; $e$, the new posterior branches of the same; B, the posterior half; $p^1$, the original proboscis; $e^1$, the original posterior branches of the intestine; $c^1$, the incipient anterior branch of the intestine; $d$, the opening of the reproductive organs. — *Original.*

branching cavities (e) became visible in the surrounding new tissue, and as they grew more distinct they could be traced along forward to the old branches (c) of the intestine. In the posterior half (B) the first step was an approximation of the anterior ends of the two lateral intestinal branches ($e^1$) toward the base of the proboscis, with which they finally made a direct, channelled communication, by projecting an outgrowth toward the central line ($c^1$) of the body, the cut edge of this segment having in the mean while, by the process of renewal, created a tissue in which to hollow these channels. As my object was merely to carry on the experiment to that extent which would instruct one in the *modus operandi* of the reproduction, I did not attempt to follow it out until I had seen the whole of the organism renewed, as that had already been done by others long before.

The point which was gained, and that was sufficient, was that in this case the regeneration was by a process of budding or direct outgrowth, precisely of the same character as that which occurs in the lower animals, which I have mentioned. Nor does this reconstruction obtain only upon a single section of the Planaria; for when cut into several pieces, from head to tail, each part will reproduce what is requisite to complete the mangled organism. And yet this is not the end of this story of marvels, for among the animals still higher than the worms, the instances of the reproduction of lost parts are none the less remarkable. The tail of a Lizard, or the legs of Crabs, Lobsters, Spiders, etc., are reproduced after being broken off. I have known a spider to leave behind seven out of its eight legs, when I put my finger on each one in succession.

Now what is *artificial division* of these animals, which in *Hydra* and *Infusoria* and *Planaria* may be carried to such an extraordinary extent without killing them? Is it not decomposition? Do we not see Hydra divided to the minutest degree, almost resolved into its original elements, as it were decomposed by the slicing operations of Trembley, and yet Phœnix-like it rises out of its own ashes?

But let us go on even nearer to the point which we are aiming at, for we can do so without drawing upon our imagination in the least. I have already pointed out the exceedingly simple structure of the Rhizopods, particularly that of Amœba (page 9). The closest examination of this animal does not reveal the least sign of a cell-like structure; it is merely a mucous or gum-like moving mass, which could not possibly be more simple unless it were resolved into a fluid condition. This creature may be divided, and even divides itself, more minutely than the Hydra allows; in fact, there is no conceivable limit to the minuteness with which it may be cut up, any more than you could imagine a limit to the minuteness with which a drop of water may be divided; and yet each subdivision moves and seizes its prey just as does the main stock from which it was separated. We could not imagine a more minute subdivision unless it were reduced to the ultimate atoms of the physicist, or actually decomposed; nor could one possibly expect it to come nearer to absolute decaying decomposition and yet retain the appearance of life.

This brings me now to the relation of some experiments, upon decomposing tissues, to ascertain what are the limits of the manifestations of life in animals. It is now five years since I first noticed the curious phenomena which I am about to describe. The experiment, or rather the discovery, was made known immediately after the investigation, and as it was written for publication whilst the subject was fresh in my mind, I will extract a part of the original article as it appeared in the proceedings of the American Academy.

"A discovery, which I made on the 20th of March, (1859,) may not be uninteresting, as it has more or less relations in its nature to the theory so earnestly advocated by Pouchet. There are certain well-known bodies described as animals by Ehrenberg, under the name of Vibrio; their peculiarity consists in that they are composed of a single row of globular bodies, resembling a string of beads, more or less curved, and move in a spiral path with great velocity, even faster than the eye can follow in many cases. They have always been spoken of as developing around decaying animal and vegetable matter. I was very much surprised to discover the manner in which they originate from such substances. I was studying the decomposing muscle of a Sagitta when I noticed large numbers of Vibrio (fig. 49, *e*, *c*,) darting hither and thither, but most frequently swarming about the muscular fibres. I was struck with the similarity of these bead-like strings to the fibrillæ of the muscle, and upon close comparison I found that the former were exactly of the same size, and had the same optical properties as the latter. Some of these appeared to be attached to the ends of the flat, ribbon-like fibres (*i*), and others at times loosened themselves and swam away. I was immediately impressed with the daring thought, that these Vibrios were the fibrillæ set loose from the fibres; but as this was a thing unheard of, and so startling, I for the time persuaded myself that they must have been accidentally attached and subsequently loosened. However, I continued my observation until I found some fibres in which the fibrillæ were in all stages of decomposition. At one end of the fibre the ultimate cellules of the fibrillæ were so closely united, that only the longitudinal and transverse striæ were visible; further along, the cellules were singly visible, and still further they had assumed a globular shape; next, the transverse rows were loosened from each other excepting at one end (*a*); and finally, those at the extreme of the fibre were agitated, and waved to and fro as if to get loose, which they did from time to time, and, assuming a curved form, (*c*, *c*,) each revolved upon its axis and swam away with amazing velocity.

" The number of ultimate cellules in a moving string varied from two to fifty; the greatest number of strings were composed of only three or four, often six to eight, and rarely as high as fifty. Very rarely the fibres split longitudinally, and in such instances the fibrillæ were most frequently long, and moved

Fig. 49. A piece of the flat, ribbon-like, muscular fibre of a *Sagitta*. 1000 diam. *i*, the fibre; *a*, fibrillæ separating from the fibre; *e*, *c*, fibrillæ separated and moving about like Vibrios. — *Original*.

about with undulations rather than a wriggling motion. A single ultimate cellule, when set loose, danced about in a zigzag manner; but whenever two were combined, the motion had a definite direction, which corresponded to the longer diameter of the duplicate combination; and if only three were combined, the spiral motion was the result of their united action. What it is that causes these cellules to move I do not profess to know, but certainly it is not because they possess life as *independent beings*. This much is settled, however, that we may have presented to us all the phenomena of life, as exhibited by the *activity* of the lowest forms of animals and plants, by the ultimate cellules of the decomposed and fetid striated muscle of a Sagitta. I do not pretend to say that everything that comes under the name of Vibrio or Spirillum is a decomposed muscle or other tissue, although I believe such will turn out to be the fact; but this much I will vouch for, that what would be declared, by competent authority, to be a living being, and accounted a certain species of Vibrio, is nothing but absolutely dead muscle." *

Just one month after this paper was presented to the Academy, I was so fortunate as to make a similar but more startling discovery in regard to the decaying conditions of another animal. The day after this discovery I communicated it to the Academy. It reads thus:—

"No longer ago than yesterday, (May 10th, 1859,) I was for-

---

* Where fibre, either muscular or tendinous, is present, it may account for the presence of Vibrios, or rather vibrio-like bodies; but when the latter occur in infusions containing simply fluids or juices of various bodies, they cannot possibly be traced to that substance. The above article, as I believe now, only proves that there are certain bodies which in a decomposing condition simulate the form and actions of Vibrios. This simply arises from the fact that all very minute bodies appear alike under certain conditions, for instance when seen with insufficient powers of the microscope. This I have more recently shown, as in the case of the decomposing muscle and tendon of the Sheep. When seen with the ordinary high powers, the vibrios of the tendons seemed identical with those of the muscle, but when more magnified with an extremely high power, the oval form of the beaded joints of the tendon-vibrio became apparent, and enabled one to distinguish them from those of the muscle when the two were mixed together.

7

tunate in discovering the origin of another, or rather of several forms of these pseudo-animate bodies called Infusoria. Whilst watching the decomposition of the inner wall of the proboscis of a young *Aurelia flavidula*, our common Jelly-fish, I observed that the whole component mass of cells was in violent agitation, each cell dancing zigzag about within the plane of the wall. If any one will shake about a single layer of shot in a flat pan he can obtain an approximate idea of the appearance of this moving mass. In a perfectly healthy condition these cells lie closely side by side, and do not move individually from place to place, but yet are active on one side, which constitutes the surface of the stomach, where they are covered by vibratile cilia. As the young Aurelia grows, this wall becomes separated from the outer one, but not completely, for the cells of the two adhere to each other by elongated processes varying in number from one to six or seven. Each cell of the inner wall contains numerous red or brown granules, a few transparent globules, and a

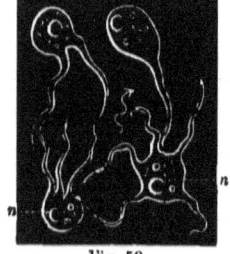

Fig. 50.

single large clear mesoblast (nucleus). When decomposition ensued, these cells became still farther separated from each other (fig. 50) and danced about in the manner which I have just described. The vibratile cilia were not observed to share in this movement; in fact I could not detect their presence, because, no doubt, they had become decomposed and fallen away; but the elongated processes, which heretofore had remained immovable and stiff, lashed about with very marked effect upon the cells to which they belonged, and caused them to change place constantly. At last the inner wall fell to pieces, and every cell moved independently and in any direction."

Since that time I have made similar experiments with the muscle and tendon of various animals, and among others those

Fig. 50. *Aurelia flavidula*. Per. and Les. 1000 diam. Cells, from the decomposed inner wall of the Ephyra, with their elongated, pointed processes lashing about with great activity. $n$, the nucleus. — *Original*.

of the Sheep; and always with the same results. In the case of the Sheep, the muscle granules were round, so that I could at once distinguish the muscle-vibrios from the tendon-vibrios, because in the latter the granules were elongate oval, even before the tendon fell to pieces in the process of decay, whilst in the former they were spherical.

Now here we see animals, some high in the scale of being, exhibiting signs of life, that is *as far as motion* may indicate it, even in the last stages of decomposition. In Aurelia, the signs of life, as exhibited by its cells, (fig. 50,) are more positive than in the other instances; but we may have various degrees of this kind of proof of vitality, according to the nature of the cells. In the case of Aurelia, the individual cells move by the action of their little arm-like projections; whereas in Sagitta and the Sheep, the granules of the vibrio-like fibrillæ may possibly move by the action of the water as it is absorbed by them, or, what is not in the least improbable, the individual granules, especially in the muscle, may contract and expand, just as they do in the living animal.

Since, therefore, we have animals whose individual cells, when separated, like so many distinct individuals, exhibit the phenomena of life, why may we not also have *one-celled animals*, existing just above the amorphous stage of decomposition? Amœba, (and Rhizopoda in general,) in one sense a one-celled animal, has scarcely a higher state of vitality than the decomposed Sheep-muscle, or Aurelia-cells. Who can say what *degree* of vitality exists in the hen's egg which is dependent, just as much as the creatures in Prof. Wyman's sealed flasks, purely upon extraneous or physical agency for its development into a chick; otherwise, if it were not supplied, or acted upon, by this physical agent, *heat*, it would remain in a low state of vitality, perhaps as low as any decomposing muscle or tendon, and finally, it would decay.

In the *egg* we have an *uncomposed* substance, and in the decaying muscle we have an animal *returning* to its former *uncomposed* state.

Therefore, if we may have, and do have, as no one doubts, an animal substance in a peculiar state, such as we find in the bird's egg, dependent upon *physical* forces for its development into an active being, is it illogical to suggest that we may possibly have some other peculiar form or state of animal substance which physical forces, in the hands of a great controlling power, can develop into life? What is digestion but a rapid method of decomposing animal (or vegetable) matter, and recomposing it directly into *living tissues?* Now, among the lower animals, there are certain kinds which, in a physiological sense, do not digest their own food. The Tape-worms which live in the various organs, stomach, intestine, liver, and brain, of animals, although some have more or less traces of a digestive organ, and others none at all, as I have shown you, (p. 83,) do not prepare their own food, but absorb it directly through the skin of the body, just as a sponge soaks up water. The food is prepared for them by the animals in which they live; and they merely appropriate the decomposed substance which surrounds them.

But it is not even necessary that the decomposed food should be prepared by digestion in other animals; for flesh which is decomposed by decay into a semi-fluid mass is absorbed by the sponge-like bodies of certain animals which live in stagnant pools. In fact, we have all possible degrees in the mode of nutrition of living beings, from that in man, in whom the digested fluid is carried from the stomach to all parts of the body in distinct tubes, the blood-vessels, down to that in the lowest animals, in which there are no vessels whatever, and in which the nutritive substance percolates through the interstices of the body.

In these lowest animals the *process* of transforming decomposed matter into their living substance is reduced to the most simple transition from death unto life that can be conceived. On the one side is the Amœba, consisting of a mere jelly-like, porous mass, and on the other side is its food, a decomposing substance, in some form scarcely less simple than itself. What

## DIGESTION AND REGENERATION.

is the nature of the transition from the one into the other, can be more easily imagined than perceived.

But now let us ask a straightforward question which very naturally arises here. Why may not different kinds of animals arise from any decomposed matter? Human digestion makes human flesh out of the decomposed meat of many different kinds of animals; the simplest form of animal life, the Amœba, likewise changes all sorts of decomposed animals into its own substance: so that at one time we may see the simple organization of the Amœba transformed into the most complicated tissues of the human structure; and at another we may find the Amœba preying upon the decaying body of man, and transforming it into the simple, crawling, jelly-like mass of its own body!

I have now only to refer you to the experiments with the sealed flasks, (page 15,) in order that you yourselves may answer the question as to whether different kinds of animals may arise from any decomposed matter. I will merely recur to one or two points of those experiments in order to reinforce the argument, and to introduce the proofs in proper sequence. You will recollect that by the manner in which the substances were introduced and *heated* in these flasks, nothing animate could by any possibility have remained in them alive. It was also shown that animals identical with those which developed in the flasks were killed by a temperature far below that of boiling; whilst in some of the flasks the heat was raised far above boiling, in one instance 38° and in another 95° above the boiling point; and yet, in these two last, as in many other flasks, living organisms were developed in abundance in the midst of the closely sealed, decomposed substance.

Now as to the nature of these living beings, which were developed in the flasks, I would say that it makes no difference whether they are animals or plants, or both. In either case it indicates the development of *life* under the sustaining influence of light, the enclosed purified air, and moisture.

One kind that was frequently developed is the ordinary *yeast-*

*plant* (page 20, fig. 7). The Bacteriums (fig. 6) are of rather doubtful nature. They may be very lowly organized animals; but I think rather that they are plants (see page 24). The Vibrios (p. 18, fig. 5, *b*) are what one might call a chain of Amœbas moving in a winding or spiral course with great rapidity. As for the Kolpoda-like bodies, (fig. 5, *c*,) no one would question that they are living creatures. They move by the action of vibratile cilia, which are eminently characteristic of life. Whether these beings develop by going through a sort of egg-state, or seed-state, we cannot say, but even if it were so, it would be none the less the development of living beings; for I have shown that the egg is merely an early stage of the animal.

And here I would say that the eggs of all animals may be said to essentially depend upon physical causes for their development; even those which are nourished by a parent for a considerable portion of their life. They live merely as parasites upon the parent, and absorb the nourishing fluids under exactly the same conditions as do the parasitic worms, which live in the liver, intestines, blood-vessels, etc., of animals. Now we may have all degrees of dependence upon the parent, from the placental absorbing relation, which I have just spoken of, to that relation in which the egg is expelled, as in birds, frogs, fishes, etc., long before there is the least trace of organs in it. Hence we have, in consequence, all degrees of dependence upon physical agents for the development of the egg into a fully organized animal.

Is it not plain that secondary causes are at work here?

Not the veriest theistic naturalist will deny, or has ever denied, that the egg, when laid, is dependent upon physical causes for development, or that those causes are secondary causes. These are facts which have become common property to all naturalists and physiologists, and yet, strange to say, many cry out in alarm if some one recognizes these self-same secondary causes where they have not been seen to apply before. Richard Owen, of the British Museum, the greatest of all the naturalists and comparative anatomists who have lived since the days of Cuvier, has been most influential in showing that secondary causes do operate more frequently than has been generally acknowledged. He argues that the development theory, in its modernized form, claims simply an extension of a well-established and long-acknowledged law, in other words an extension of the *degree* of prevalence of this *law*. He propounds no new rule, but in a straightforward manner argues that the Creator works according to *recognizable laws* of his own establishing.

It does not follow, that, as beings may originate through secondary causes, the Creator is any the less connected with the causation than he would be by a direct fiat; it simply means, as I have already said, that the Creator makes the processes of his work visible by the nature of the laws which he has established to reign among all things.

How is it that we estimate the intellectual standard of mankind? Do we not more highly regard the man who invents a machine to mow with, than him who cuts his grass with a common scythe; or him who makes a steam-power printing-press, than the one who prints with a hand-press? Do we not, all of us, consider the invention of any sort of apparatus, to do that work which has heretofore been done by hand, as an indication of a higher order of mind than that possessed by the manual laborer? What do not the inventions of this war tell for us as a nation? In fact, the production of any work by machinery is a production by secondary causes; but none the less under the control of man. Now shall we fail, then, to recognize that Creator, who works by a *method* which we can see and understand, as a higher order of being than one who works in ways mysterious to us all? The latter is the god of the savage, of the barbarian; the pantheistic god; the god among other gods; the god subordinate to some preëxisting god; and the last, the creature and creation of still another, long before him; and this from yet another, and another, and another, ever and eternally preëxisting the last; but the God of the intelligent man is a power within himself, — *self-existent*, — *a Unity!*

It by no means follows that because we claim that the Creator works according to an order of things, by a method, or law, and that his existence is involved in, or implied from the evidence of a plainly operating law, which must of necessity, to our minds, have an intelligent thought and power to guide it, I say it by no means follows, from this idea, that the Pantheist can step in to claim that the law-giving Creator must have derived his law from a previously existing idea of law, which he received from a still higher power, and that that power received it from one yet beyond; such a thought kills itself, for its demands have no end, and leaves its pursuer groping about in a mist of infinities. The one God, that infinity of power encircled and concentrated within himself, the God of reasoning, intelligent man, is the more and more highly estimated and reverenced as his works are the better understood; by so much as every newly discovered law, or the extension of one already known, helps to explain the so-called mysteries of nature.

The science of Theology is not now what it was only a few years ago. How largely Natural History is at present drawn upon by the theologian, not only to prove the existence of a Supreme Being, but even to show the nature of his moral qualities, his goodness, his benevolence, and his power! Year after year Theology continues to add to its store even those facts of science which have at one time been looked upon as dangerous to the faith.

Geology was in former times shunned as the worst form of atheism, because it taught that there were animals, buried in the earth, which lived at periods long anterior to the creation of man and the hosts of creatures which crowded about Adam in the garden of Eden. But now how changed. We have seen one of the most orthodox of theologians, Hugh Miller,— peace to his ashes!— writing whole volumes to prove the existence of animals, with a backbone, as far anterior to man's creation as possible! Paley's " Natural Theology " abounds in instances of secondary causes. This was and is the natural history textbook of the theologian. I do not think it is difficult to see which way the current runs in this matter.

## THE PRIMORDIAL STATE OF ANIMALS. 105

But now let us see what would be the natural consequence of some of the arguments which have been urged *pro et contra*. Although it may be admitted that all animals, except such as arise by budding, reproduce themselves by eggs, yet it does not follow that the first created animal, like its successors, was an egg; for to assert that would be to assume that eggs are not or have not always been uniformly developed in a parent, since the first egg could not then have had a parent; but if it were true, then, it would appear that some eggs have been altogether dependent upon physical causes for their origin, at least for their surroundings or matrix; and if so *once*, why not *again:* why should there be a change in the process of the original creation?

Again, if, on the other hand, it is assumed that the first animals were created full standing adults, then it is admitted that some animals have originated without passing through the egg-stage, and thus the probability is left open that more and other kinds of animals have been created in the same way since the beginning; and hence it follows, of necessity, that the advocate of this theory must admit that spontaneous generation, that is, creation without the preliminary inter-parental egg-state, is possible. How, or by what causes, whether the primary creative or secondary, it matters not; the question here is, does it happen? The essential point to be ascertained in such as Prof. Wyman's flask experiments is, having so disposed matter in the apparatus that one is sure that no living thing exists there, whether animate beings do originate there; and if they do, it must be without the intervention of any previously existing animal; that is, they must be created there either by direct fiat or according to the laws of the Creator. The theory of *spontaneous generation,* as a fact, has nothing to do with the *how* it is brought about; but simply *does it occur;* do individual animals ever originate totally independent of other individuals?

Now if the opposing theorists advocate the egg theory, then they must admit that there has been a want of uniformity in the mode of the creation of animals; seeing that eggs at one time originate without a parental matrix, and at another time within a parent, *i. e.*, under two totally different circumstances, according to their ideas. If on the other hand they claim that the adult stage was that in which the first animals were created, then, as I have said before, they admit that spontaneous generation has happened at some time or other. If so, then what is there in the theory of spontaneous generation so revolting to the theistic creationists, or that it should have happened since the first time, — that the Creator should have thought to repeat and continue his original plan of creation, as one of several modes of giving rise to individuals?

No one revolts now at the statement, the fact being well established, that large groups of animals arise by budding; that is, by a division of one individual into two or more parts, and each part becoming a perfect animal. This is a stage which has no egg-phase, and no one disputes its occurrence; yet before it was verified, must not the idea have seemed just as absurd as some would have that of spontaneous generation? The latter only differs now in probability from the former in that it lacks so palpable and as numerous proofs as that. The restoration of lost parts, such as legs, jaws, and tails, by animals, from the highest to the lowest, with increasing degree of frequence as we go lower, is akin to budding, and makes budding seem not impossible even in the highest animals. Yet that a whole limb or several limbs of a crab, lobster, spider, caterpillar, or grub, the arms of a starfish, or the tail of a lizard, are restored after having been broken off, was once as marvellous as spontaneous generation now is.

It would seem to be more plausible that the *adult* phase should have been the *primordial* state of the first animals; simply because in that state animals could have taken care of themselves, whereas eggs would have been at the mercy of the elements, and that would not accord with the known condition of eggs in their first stages. To have the theory regarding this dependent form consistent with itself, the egg must have always originated in a parent. The only other alternative is that the egg-phase is not a distinct one from that of budding. This again would add still more to the probabilities of the occurrence of spontaneous generation, because it reduces the egg from a distinct, as it were an individualistic feature, to one which is merely a passing characteristic in the life of an animal. It then loses caste, its *potentiality* is gone, it passes out of a high controlling state, which conditionates and precedes all subsequent phases, to a subordinate state, which it then must hold in common with, and simply as one of the first of several sequences, which are all of equal value, at least inasmuch as one of them cannot be omitted without destroying the continuity of the successive phases characteristic of the life of a living being. In this subordinate state, then, it cannot be claimed for the egg that *its* peculiar, idiosyncratic condition should be the one any more than any other, for instance such as budding, which originally began the animal phase of being on earth; for although the budding phase implies necessarily an individual to bud from, yet it is hardly more clearly related to some previous individual existence than is the egg. The physiological and normal relation of the egg is to be evolved from a reproductive organ, an ovary, and an ovary implies an animal to possess it. This could all happen according to the theory of spontaneous generation, for in that case the adult animal is formed to preëxist the egg. Origin by budding, and by eggs, ovogenesis as it is called, being therefore put out of the question, the last mode left is that by spontaneous generation, which essentially amounts to a repetition of the great original, primary act of the direct creation of animals in an adult state.*

Thus the matter stands in my mind. If I may not appear to be right in the eyes of some, pray tell me who has been inspired with the revelation of the truth, for to him alone would I listen with deference ; otherwise the question is open alike to the reason of every mind, without any preponderance in favor of the

* I use the term "*adult state*," not as strictly meaning *full grown*, but rather as indicating any of those *free conditions* of an animal, above the fixed, *confined* state of the *egg-phase*.

authoritative *dictum* of any one more than another. Therefore I say that it stands to reason that secondary causes are the visible modes of the action of the Creator's will, and that his great primary fiat has not ceased to exert its influence even at the present day.

## CHAPTER V.

SPONTANEOUS GENERATION AND REPRODUCTION BY BUDDING AND FISSIGEMMATION MOST FREQUENT AMONG THE LOWEST RANKS OF ANIMATE BEINGS.— ALL ANIMALS ALIKE IN THE EARLIEST STAGES.— MAN AND MONAD ARE AT ONE TIME A MERE DROP OF FLUID.

No doubt the first question that will arise in the minds of some in regard to spontaneous generation is this, " If spontaneous generation is truly one of the modes of introducing life on earth, why do we not see animals of the higher forms and grade spring into life, as we do among the lower ranks ? " To this I would answer, in the first place, by asking another question, namely, Why do not the higher animals, such as the quadrupeds and man, reproduce themselves by budding, or by transverse division, as frequently, and as a *rule*, as do the lower animals ? Now in regard to this, I pointed out the other night that what is a rule among the lower animals is an exception among the higher forms of life, and that, when budding or self-division does occur among the latter, it is under the guise of what is called Monstrosity, which I explained as an abnormal recurrence to the lower modes of reproduction, consequent upon a low state of vitality in disease, or hereditary degeneracy.

I think, upon due consideration, it will be found that reproduction by eggs is a common characteristic of all animals, whereas budding or self-division is, within a certain limit, a peculiarity which does not extend to the higher forms. We find budding or self-division occurs in the lowest groups of all the animal types. Among Protozoa, as the Infusoria are now called, budding and self-division have been thought to be the only modes of reproduction, until within a few years, when it was discovered that they also reproduce themselves by eggs. Among Zoöphytes the Corals form their branching, tree-like stems by budding. In the group of Mollusca, (shell-fish,) the Bryozoa, which resemble so much, and are often mistaken for those beautiful and variously-colored sea-weeds which we meet with everywhere on our coast, form their graceful branchlets and plumes by budding. Among Articulates, the aquatic worms multiply by self-division, which, in some instances at least, is a budding process also. The intestinal worms multiply at an enormous rate by self-division. Even among Vertebrates, it has been found that the lowest of them, the Fishes, also very frequently, in their embryonic state, self-divide, as Lereboullet's observations have so abundantly proved (see page 85).

These characteristics are those which, as it were, reign among the lowest forms of all groups; and there is no more reason why spontaneous generation should occur as a rule to produce the higher animals, than that budding and self-division should do so. Neither in the one case nor the other do we know why these peculiarities are excluded from the higher groups. We can only perceive that the phenomena are connected with some, as yet but partially explained, law of development and derivative succession. When Trembley, in 1744, announced the wonders of the budding Hydras, and their regeneration after they had been cut into numerous pieces, the question might have been asked, with equal reason, by the skeptic, why do we not see man and the quadrupeds budding, and reproducing their arms and legs when cut off?

Spontaneous generation, I think, stands in the same relation to the higher forms that budding and self-division do; it is a phenomenon confined to the lower forms, but extends upwards more or less, and in various conditions; — thus, certain of the Protozoa originate by spontaneous generation, *i. e.*, altogether independent of a parent, and rely upon secondary causes for their development; certain other animals are partially dependent upon a parent, but for the greater part of their embryonic growth they depend upon secondary causes for the completion of their development; and these various degrees of independence of a parental influence we have seen correspond in certain ways with the rank of the parent.

## EARLIEST PERIOD OF GROWTH. 111

There is one fact, moreover, that must not be forgotten; it is this, that *all animals, from the monad, the gum-drop Amœba, up to man, at one time cannot possibly be distinguished from one another!* I mean in the *earliest egg-state.* When I described to you the other night the manner of development of the egg, I showed that all eggs (page 33) commence as a mere *drop of fluid,* with certain bipolar characteristics. Now whether this drop of fluid be taken from the Amœba, the coral, the worm, the shellfish, or the rabbit, you could not tell the one from the other, any more readily than you could distinguish a drop of water from Cochituate Lake from that of Mystic River! The distinguishable characteristics appear in the process of time and development.

Now, when I say that all animals, from the monad up to man, at one time cannot possibly be distinguished, *per se*, from one another, I mean to draw your attention to *another* fact, which is, that *all animals at one time are as simple as those which are thus far known to arise by spontaneous generation.* The higher animals do not at once leap into life in a full-grown state, but they arise out of as simple elements as do those lowest forms which originate through spontaneous generation. The higher forms pass through the same *relative conditions* as do the lower ones, and then rise to higher states. This idea is also carried out in another way, in restricted groups, as I have shown you in a previous lecture (p. 40); thus the quadruped, when in its earlier stages, possesses a conformation of its principal organs — that is, the heart and nervous system — similar to that of a fish; then it takes on the form and relations of the next higher group, namely, the reptiles, and then that of birds, and finally its adult conformation.

In the instance of spontaneous generation, the animals commence these simple stages without a parent; whereas, among higher animals, their simple drop-of-fluid stage is farther removed from external causes, and is subjected to parental influence as a primary cause, and secondary causes affect the progeny through the parent at first, and afterwards directly, when the egg-animal is cast forth upon the world. In the latter case the action of secondary causes is complicated in correspondence with the rank of the animal. The prevalence of secondary causes is a matter of degree, from the highest degree of direct action to the lowest or least degree of indirect influence.

Another feature is not by any means to be overlooked, namely, that the quality or intensity of the influence of secondary causes varies not only with the rank of the animal, but also, I am strongly inclined to believe, even in the same species, according to the varying high or low degree of vitality of the creature. Here I refer you again to the case of the monstrosities.

I am well aware that it is a difficulty with some to comprehend how we creatures, who are so "fearfully and wonderfully made," who embrace in ourselves such a variety of forms, relations, proportions, and properties, should be in any way subject to the action of secondary causes; but if I were to refer such persons to the chemist and physicist for the properties of crystals, they would be told that there is as remarkable definiteness in their geometrical forms, each corresponding to its peculiar chemical character, whether it is salt, or soda, or alum, or sulphur, or iron, or gold, or quartz; that each has its own mode of action upon the rays of light, and also certain relations to magnetism; in fine, that these *inanimate* crystals have as definite laws as have *animate* beings to rule among them;— and that yet, with all their varied properties, they are *totally subject* to *secondary causes.* Had I time to enter into the subject, I might show you how all the animate and inanimate forms, animals and plants and minerals and fluids and gases, &c., are related in certain of their features or characteristics to geometrical forms; and how mathematicians have demonstrated that geometrical forms are merely the more tangible expressions of great mathematical laws, which rule the movements of this mighty universe! But I must refrain from any further consideration of this immediate subject, lest we should become too far diverted from the main point in question.

We have seen thus far that all the varied phenomena concerned in the production of animals, are based upon the idea of *degree*. We have the degree of rank among animals, from the lowest to the highest. There is the degree of complication in their eggs. Then the degree of prevalence of budding and self-division. The degree of individuality. The degrees through which the higher forms pass to reach their adult state. The degree of relation to the influence of secondary causes, and the converse, the degree of relation to the parental influence. The degree of intensity in the action of secondary causes corresponding to the degree of vital power in the same species. And now I will add, but not dwell upon for the present, that the degree of rank in which animals have first appeared, in past geological ages, is a low one for all groups.

Thus you see all is a matter of *degree*, — of *progression*, from a lower to a higher state; and consequently we should very naturally suppose that the phenomenon of spontaneous generation would make itself most conspicuous among the lowest forms, when it occurred at the present day. It is not possible, in the existing state of science, to say to what extent, or through how many groups of animals, spontaneous generation prevails. The difficulty of imitating the processes of nature, and isolating them, as has been done in the case of the sealed flasks, renders the multiplication of such observations a matter of exceedingly slow progress. Having gone now as far as the limits will allow in the investigation of the origin of the first and lowest stages of life, we come next to the consideration of the relation of the succeeding, more elevated phases to the latter.

# PART SECOND.

## THE FIVE GREAT ANIMAL GROUPS.

# PART SECOND.

## THE FIVE GREAT ANIMAL GROUPS.

### CHAPTER VI.

THE IDEAL TYPES.—ALL ANIMALS BILATERAL.

WHAT I have said thus far did not especially involve the discussion of the relative rank or grade of animals as a whole; the point to determine was whether animals ever develop as I have stated. We come now to the discussion of the question as to whether the higher animals have any relation to the lower ones, and what those relations are. Are they relations of mere similarity, simply because they all exhibit life? Are they such relations as would indicate that the higher have developed from the lower, or *vice versa*? Or are they such relations as would seem to show that they have affinities with each other by groups? All naturalists admit that they are allied to each other by groups, in some form or other.

Lamarck arranged animals in two principal groups, namely: the *Vertebrata*, or such as have a backbone or vertebral column; and the *Invertebrata*, or such as have no backbone, but have, as he says, a skeleton on the outside. Cuvier considered the Animal Kingdom to be divided into four groups. Vertebrata, Mollusca, Articulata, Zoöphyta. The last three correspond to Lamarck's Invertebrata.

Lamarck believed the classes of his groups to be genetically related to each other; and he constructed a tabulation of the various groups, which should exhibit in what way, or through what lines or passages, these minor groups most probably have run in order to form a connection with each other. On the other hand, Cuvier seemed to believe, (I say seemed to believe, because he speaks of certain groups, the Barnacles for instance, as establishing, "by several relations, a sort of intermediary between" the Mollusca and the Articulata,) his four groups to be four distinct *types*, or *ideas*.

Thus, he says, the Vertebrata are principally characterized by having "the brain (fig. 51, *en, cr*) and the principal trunk (*nr*) of

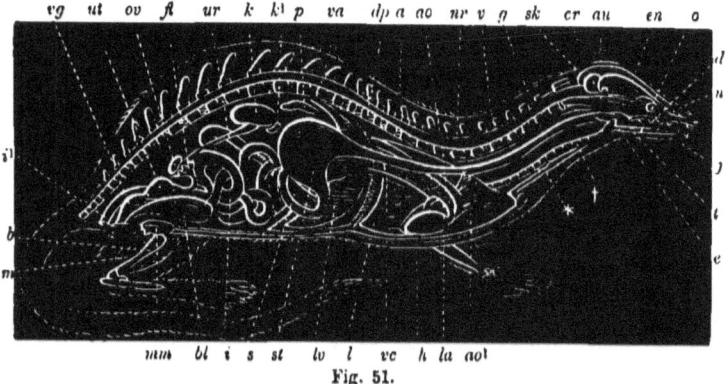

Fig. 51.

the nervous system enclosed in a bony envelop, which consists of a skull (*sk*) and a vertebral column (*v, va*), or backbone."

The Mollusca, commonly called shell-fish, " have no skeleton; the nervous system, which consists of several scattered masses

Fig. 51. A diagramic longitudinal section of a Mammal. *sk*, skull; *v*, vertebræ; *a*, dorsal arches of the vertebræ; *va*, the upper and lower portions of the vertebral arch; *j*, lower jaw; *b*, bone of the leg; *m*, muscle; *d*, teeth; *t*, tongue; *g*, gullet; *, thyroid gland; *st*, stomach; *i*, intestine; *i*¹, end of *i*; *lv*, liver; *p*, pancreas; *s*, spleen; *k*, kidneys; *k*¹, appendages to *k*, known as the suprarenal capsules; *ur*, outlet of *k*; *bl*, bladder; *e*, epiglottis, or entrance to the windpipe (†); *l*, lung; *h*, heart; *ao*, abdominal aorta; *ao*¹, carotid artery going to the head; *vc*, vena cava inferior, or abdominal vein; *la*, pulmonic artery; *dp*, diaphragm; *o*, the eye; *en*, cerebrum; *cr*, cerebellum; *n*, olfactory nerve; *au*, the outer ear; *nr*, spinal marrow, or main nervous cord; *ov*, the ovary, or eggbearing portion of the reproductive organ; *fl*, the trumpet-shaped Fallopian tube through which the eggs pass into the uterus (*ut*); *vg*, the vagina, or outlet of *ut;* *mm*, the mammæ, or milk-bag. — *From Owen. Slightly altered.*

(fig. 52, $sg$, $g$, $og$) united by nervous threads ($c$, $n$), and the intestines are all included within one general uniform envelope."

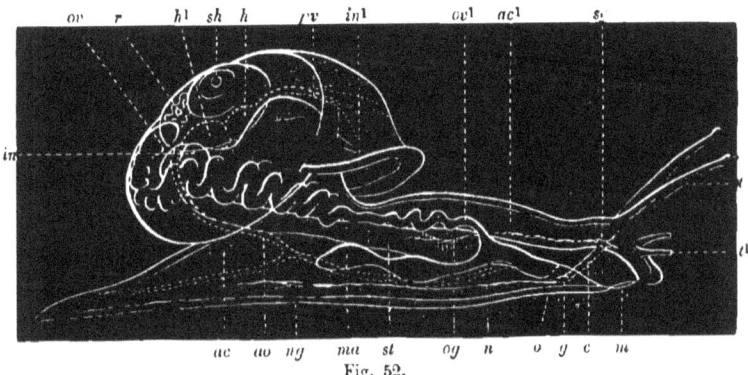

Fig. 52.

The Articulata, which include the jointed animals, such as insects, spiders, crabs, earth-worms, leeches, &c., he says,

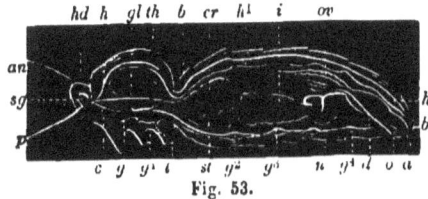

Fig. 53.

" have a nervous system, consisting of a double cord (fig. 53, $c$, $n$) which runs along the lower side of the body, and is swollen from space to space into knots, or ganglions, ($g$, $g^1$, $g^2$, $g^3$, $g^4$). The envelope of the body

Fig. 52. *Helix albolabris.* Diagramic representation of the common Snail. 2 diam. $ac$, $ac^1$, the abdominal cavity; $sh$, the shell; $t$, the larger pair of feelers, with an eye ($e$) at the tip of each; $t^1$, the smaller pair of feelers; $m$, mouth; $st$, stomach; $in$, intestine; $in^1$, posterior opening of $in$; $sg$, the superior ganglion of the head; $g$, the inferior, or sub-œsophageal ganglion; $c$, the nervous collar; $n$, $ng$, the foot nerves; $og$, the œsophageal, or gullet nerves; $h$, the auricle, and $h^1$ the ventricle of the heart; $ao$, the aorta, or main artery; $pv$, vein from the lung, or pulmonic vein; $ov$, the ovary, or egg-bearing organ; $ov^1$, the oviduct, or emptying conduit of $ov$; $o$, exterior aperture of $ov^1$; $r$, the fertilizing gland, or male element of the reproductive organs; $ma$, the matrix. — *Original.*

Fig. 53. *Sphinx Ligustri.* Lin. The Privet Hawk-Moth. Natural size. A longitudinal, sectional view. $an$, antennæ, or feelers; $hd$, the head, or first joint of the body; $th$, the thorax, consisting of the 2d, 3d, and 4th rings; $b$ to $b^1$, the eight rings of the abdomen; $l$, the base of the legs; $p$, the tubular proboscis; $gl$, the gullet; $st$, stomach; $cr$, crop; $i$, intestine; $a$, posterior end of $i$; $h$, $h^1$, $h^2$,

is divided by transverse folds into a certain number of rings, ($th$, $b$ to $b^1$)."

In his fourth grand division, Zoöphyta, — or Radiates, as he frequently calls them, — in which he includes not only starfishes, sea-urchins, jelly-fishes, and corals, but also intestinal worms, and infusoria, "the organs are arranged like rays around a centre," or, as he expresses it in one place, (vol. III. p. 218, ed. 1829–30,) "along two or more lines going from one pole to the other (fig. 54)."

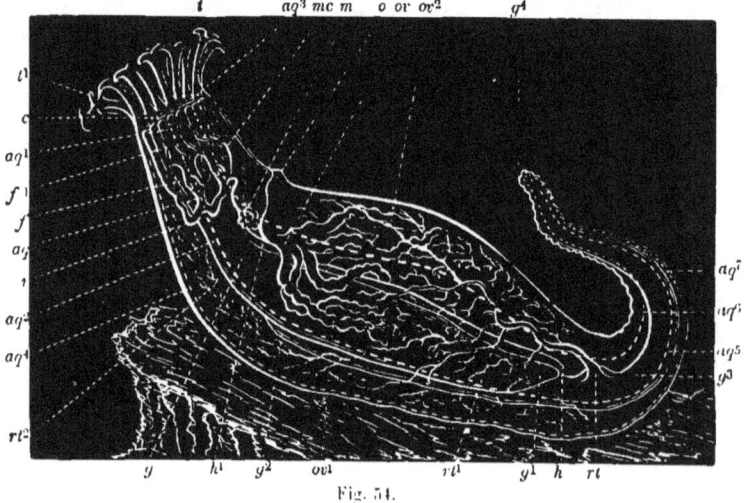

Fig. 54.

As a general thing, naturalists have accepted the divisions of Cuvier; but there is a diversity of opinion in regard to the limits of these divisions. Some accept them in the same sense as

the heart; $sg$, superior nerve ganglion of head; $g$, $g^1$, ganglions of the thorax; $c$, nervous collar; $n$, main abdominal nerve; $g^2$, $g^3$, $g^4$, ganglions of $n$; $ov$, ovary; $d$, oviduct; $o$, exterior aperture of $d$. — From Newport. Slightly altered.

Fig. 54. Caudina arenata. Stmp. Natural size. A longitudinal, semi-diagramic view of a common Trepang of our coast. $t$, $t^1$, the four-pronged, anchor-shaped feelers of the head; $f, f^1$, the stave-like, calcareous, forked pieces of the buccal ring; $g$, the anterior end of the intestine; $g^1$, the first bend of the same; $g^2$, the second bend of the same; $g^3$, the posterior or cloacal region of the intestine; $g^4$, posterior aperture of the same; $rt$, $rt^1$, $rt^2$, the respiratory branches; $m$, the madreporic body; $mc$, the madreporic canal; $r$, the aquiferous ring; $aq$, the aquiferous canals going from $r$ to the space ($aq^3$) at the base of

Cuvier presented them, or even in a more stringent sense, making each grand division an absolute circumscription within itself. Others look upon these four groups, or rather *five* groups, — since the Protozoa (Infusoria) are regarded at the present day as a distinct division from the Zoöphyta, — as so many subdivisions which have more or less intimate relations with each other; as if they were the components of a vast cloud which is divided into five masses, having their edges mutually merged into each other.

In a certain accordance with this idea of the mutual relations of the five divisions of the animal kingdom, I have constructed these diagrams (figs. 55 to 64), to illustrate the corresponding positions of the organs in the typical forms.*

the feelers (*t*); $aq^1$, $aq^2$, $aq^4$, $aq^5$, $aq^6$, $aq^7$, the longitudinal aquiferous canals running close to the under surface of the skin; $h$, $h^1$, the heart; $c$, the ribbon-like nervous collar; $ov$, $ov^1$, $ov^2$, the reproductive organ; $o$, the external aperture of $ov$. — *Original*.

* It may be objected here that some of the animals in these diagrams are placed upside down, in order to bring the organs into corresponding position in all of them; but I would ask, then, do animals have any definite relation to up and down? Which, for instance, is the back among the Zoöphytes? The *Holothurians* (Trepangs, fig. 54) creep on the side exactly opposite to that on which the Sea-urchins do! The latter creep in the position which the diagrams (figs. 57, 58,) represent, *i. e.*, *heart downwards*. Among the Mollusca, the Cuttle-fish and Squid (ch. XI., fig. 124) swim usually backwards, and with the *back downwards*. Now these comprise a large group among Mollusca. Another considerable group among Mollusca, the so-called Nucleobranchiata, allied to Conch-shells, swim rapidly in the sea *back downwards*. Many kinds allied to the clams live in holes in the sand and mud, with the *head downwards*, and others, like the oyster, rest on the side!

Among Articulata, who can say which is the *back* of those intestinal worms (the Tape-worms, fig. 44) which live on the juices of the stomach of various animals? Even among the Insects, in which back and front seem to be most distinctly marked, the Notonectas, water-bugs, invariably swim *back downwards*. Of the Vertebrates, the Halibuts and Flounders creep and swim on the side. The Bats rest all day hanging by their hind legs, head downwards. The Sloths, a curious group of quadrupeds, always move among the branches of the trees, where they constantly live, *back downwards*, hanging from the limbs by their long claws. Now if it should be urged that the Sloths nevertheless have what is to them their *terra firma* next the lower side of the body, we would ask, why

122                THE IDEAL TYPES OF THE

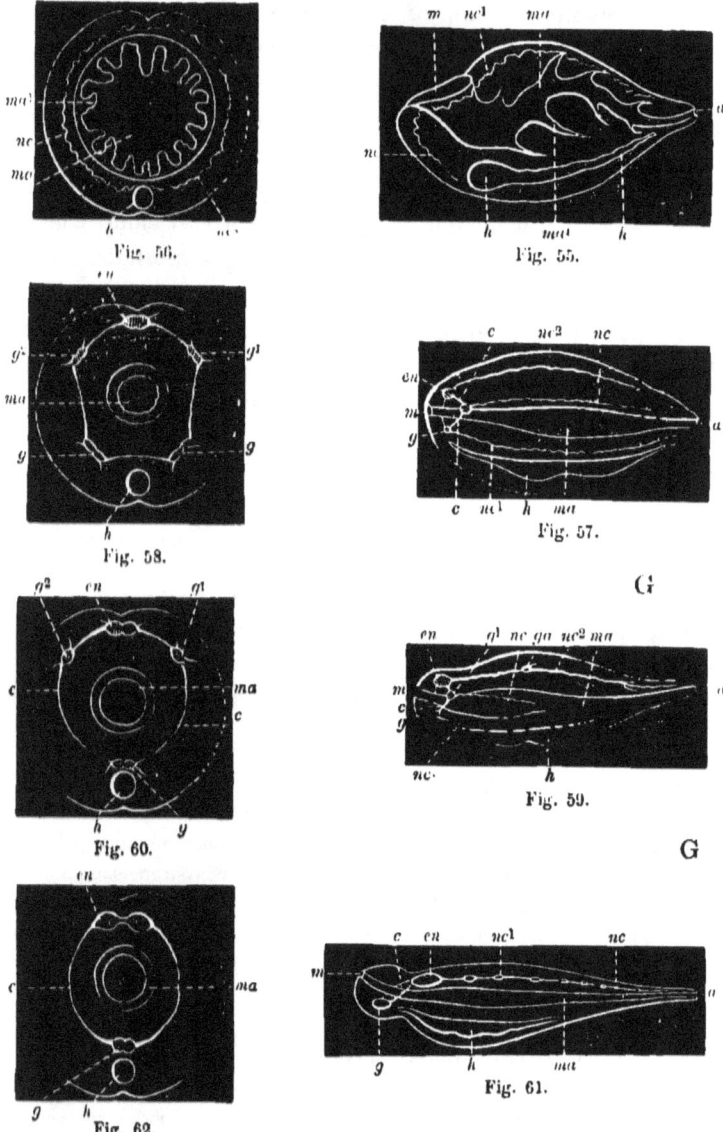

Fig. 56.
Fig. 55.
Fig. 58.
Fig. 57.
Fig. 60.
Fig. 59.
Fig. 62.
Fig. 61.

then does the Notonecta invariably keep his back in one direction, *i. e.*, downwards, when his *terra firma*, the water, is all around him? This is enough to show that the matter of up and down has nothing whatever to do with the relative position of the organs of animals.

## FIVE GREAT ANIMAL GROUPS. 123

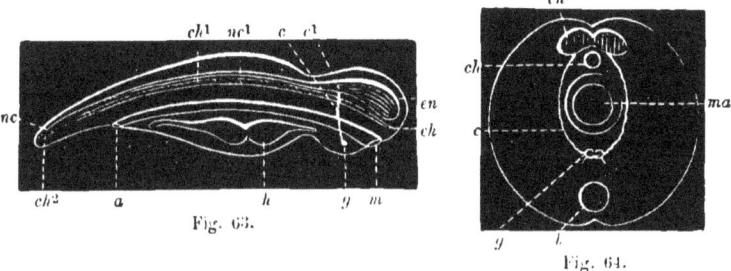

Fig. 63.

Fig. 64.

Now if you will call to mind what I have said in regard to the nature of the connection of these groups with each other, when I compared them to clouds which are more or less merged into each other at their margins, you will understand me when I tell you that these diagrams not only represent the *type* or *symbolical form* of each group, but also exemplify the tendency of one group to merge into another. What the meaning of this tendency is, I will make apparent in good time.

Thus, the Protozoa (figs. 55, 56) have an irregular, hardly defined, digestive cavity, ($m$, $ma$, $ma^1$, $a$,) on one side of which is a hollow, ($h$, $h$,) which beats like a *heart*. The *nervous system*, according to the most recent researches, (see p. 64,) is a layer of *nerve-cells*, or threads, ($nc$, $nc^1$,) which lie just beneath the surface of the body.

In Zoöphyta, (figs. 57, 58,) the intestine ($m$, $ma$, $a$) is more evident as a canal than in Protozoa, and, among the highest, the *heart* ($h$) is a distinct tube which runs along one side of the body, whilst the nervous system, ($en$, $g$, $g^1$, $g^2$, $c$, $nc$, $nc^1$, $nc^2$,) although disposed around the body in an apparently diffuse manner, has its principal parts so regularly posited as to stand in *perfect symmetry* with the other organs; thus it has two ganglions, ($g$, $g$,) one on each side of the heart, ($h$,) one also on each side ($g^1$, $g^2$) *above*, and one directly on the *median line*, ($en$,)

Figs. 55, 56, longitudinal and foreshortened views of an ideal *Protozoan*; figs. 57, 58, the same of a *Zoöphyte*; figs. 59, 60, the same of a *Molluscan*; figs. 61, 62, the same of an *Articulate*; figs. 63, 64, the same of a *Vertebrate*. The corresponding parts of the organization are lettered alike in all. — *Original.*

all of which are united by nervous threads ($c$) into a ring, which is often called the *nervous collar*.

In *Mollusca*, (figs. 59, 60,) there is, as before, a distinct intestine ($m$, $ma$, $a$) in the centre, but the heart ($h$), as a general thing, is more concentrated than that of the Zoöphyta, and the nervous system ($en$, $g$, $g^1$, $g^2$, $c$, $ga$, $nc$, $nc^1$, $nc^2$) *preponderates* on the side *opposite to the heart* ($h$), whilst there is one or more ganglions ($g$) next to it. All of these ganglions are united by nervous threads ($c$) into an irregular circle about the body. The Mollusca were considered by Cuvier as next in rank to the highest animals, the Vertebrata; and such is the opinion of the most eminent of all his successors, Richard Owen. The reason is obvious; for the Mollusca, especially that group of them which comprises the Cuttle-fishes and Argonautas and Nautilus, have an organization which in part is more nearly related, in complicity and kind, to the Vertebrates, than is that of any of the other grand divisions. I have said that the Molluscan organization is, *in part*, superior to that of all others below Vertebrates; for anatomical investigations of later years, and observations upon the intelligence of Insects, have led many naturalists to look upon Articulata as parallel with Mollusca in point of rank. Let us see what is the tendency among them as regards the nervous system, the ruling power of life.

In *Articulata*, (figs. 61, 62,) the intestine, ($m$, $ma$, $a$,) still in the centre, is bordered on one side by the heart, ($h$,) as in previous groups, and on the opposite side we find the nervous system, ($en$, $g$, $c$, $nc$, $nc^1$,) nearly altogether concentrated along a median line, and showing a strong advance toward the head, but on the side next the heart ($h$) a single ganglion ($g$) which is united to the main group by a nervous ring ($c$). The tendency to concentrate the *life-giving system* toward the head is illustrated by the longitudinal section.

In the Vertebrata, (figs. 63, 64,) we have the highest degree of concentration of the nervous system, ($en$, $g$, $c$, $c^1$, $nc$, $nc^1$). The main group of nerves is massed as a single chain, ($en$, $nc$,) — still with traces of a double character, as in Articulata, — and this chain is on one side of the intestine, *opposite to the heart* (*h*), but yet one of the ganglions, that of the sense of taste, (*g*,) one of the chief senses, is placed on the opposite side of the intestine, and *next the heart*. The position of the internal skeleton is represented by a line (*ch, ch$^1$, ch$^2$*) running between the main nerve-trunk (*en, nc$^1$*) and the intestine (*m, a*).

I know some naturalists would object that the Mollusca, Articulata, and Zoöphyta cannot correspond, in the nervous system, with the Vertebrata, because the first three have a nervous ring or collar, (*c*,) and the brain is on the opposite side to the nervous cord; but this, upon careful comparison, will be found to be not true; for, in the first place, the Vertebrata do have as distinct a nervous collar (figs. 63, 64, *c*) as some Mollusca, and secondly, one of the principal senses, *taste*, is centred in a ganglion (*g*) which is reached from the main nerve by this *collar* of nerve-threads which passes around the intestine exactly in the same way as in Mollusca and Articulata. Moreover, in regard to the position of the so-called brain in Mollusca and Articulata not corresponding to that in Vertebrata, I would say that in the Mollusca the position of the sense-ganglions varies, but in the highest of them, the Cuttle-fishes, the ganglion for the eye holds almost the same relation to the main trunk as does the corresponding organ in Vertebrata; and in the Articulata, the position of the *eye-ganglion* also varies, for although among Insects it is on the opposite side from the main nerve, yet among a large group of Worms, the Nemertians, the ganglion from which the optic nerve springs is on the same side as in the Vertebrates. *Between these two extremes of position there are all grades.*

As to the objection raised by those who claim that the vertebral column, or *bone* of the back, is the distinguishing feature of Vertebrata, I would say that when we refer to certain fishes, such as Lamprey eels and the Sturgeons, we find the backbone represented by a mere *gristly cord*, and in other fishes (Myxine, Ammocœtes) a *cord* of *interlaced fibres* in place of vertebræ. Yet in several respects these fishes are far more highly organized than the so-called bony-fishes; their brain is of a higher order; their circulatory and respiratory system is also superior; and the organs of reproduction are not only higher, but approach closely to those of reptiles. So we see that the *bony nature* of the vertebræ is not the essential characteristic of Vertebrates; it is the *presence* of some longitudinal mass, either bone, or gristle, or interlaced fibres, or a mere jelly-like string, as in *Amphioxus*, (fig.

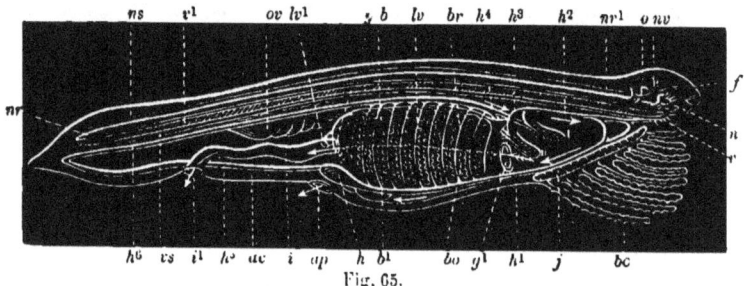

Fig. 65.

65, $v$, $v^1$,) which is intended to separate the main nervous cord from the rest of the organs; it is the *presence* of this *ideal*, as I may call it, that constitutes the characteristic of Vertebrates. In other words, it is an *ideal* axis *materialized*.

From this you may judge that in considering *typical* forms of life, it is the *relation* and not the *nature* of a substance which is to be taken into account. Relation should be the ruling standard.

Accordingly, therefore, we see in the Protozoa, (figs. 55, 56,) the

Fig. 65. *Amphioxus lanceolatus.* A diagramic figure of the Lancelet. Natural size. $f$, the head; $r$, $v^1$, the notochord, or vertebral column; $vs$, the sheath of $v$, $v^1$; $bc$, the buccal cirrhi; $j$, the buccal ring at the entrance to the mouth; I, II, oval bodies projecting freely into the buccal cavity; $g^1$, entrance to the throat or branchial cavity; $g$, posterior end of the same, and entrance to the intestine proper ($i$); $bo$, the lateral branchial openings; $i^1$, posterior end of $i$; $lv$, $lv^1$, appendage to $i$, opening into it at $lv^1$; $h$, the heart; $h^1$, $h^2$, the anterior blood-vessels; $h^3$, branches from $h^4$, supplying I, II; $h^4$, $h^6$, the dorsal artery; $h^5$, the abdominal vessel; $b$, the upper, and $b^1$, the lower point of junction of the branchial vessels ($br$) with the dorsal ($h^4$, $h^6$) and ventral ($h^1$, $h^5$) vessels; $ac$, abdominal cavity; $ap$, abdominal pore; $nr^1$, the anterior, and $nr$, the posterior end of the main nerve or spinal marrow; $ns$, sheath of $nr$, $nr^1$; $o$, the eye; $n$, the olfactory nerve; $nc$, the facial nerves; $oc$, the reproductive organ. — *From Owen.*

nervous system, although formless, holding a *certain position*
in reference to the other organs. In Zoöphyta, (figs. 57, 58,) it
is more collected, and arranged symmetrically, in reference to
right and left, and above and below. In Mollusca, (figs. 59, 60,)
it is more concentrated toward the side opposite the heart. In
Articulata, (figs. 61, 62,) the concentration is still further carried
out; and finally in Vertebrata, (figs. 63, 64,) the nervous system
attains its highest confluence not only toward the median line
opposite the heart, but also in its tendency toward a *head*.

Now it is a remarkable fact, that, as we trace the arrangement
of the systems of organs from the lowest to the highest groups,
we find the tendency is, in one sense, to carry out the idea of
*polarity*, which we see in the egg, to its strongest expression.
Thus, among the lower animals, the egg has the two poles (page
34, fig. 15) not distinct, since the opposing oil (*ol*) and albumen
(*alb*) merge more or less into each other; but among the higher,
(page 35, fig. 18,) the two poles, *yolk* (*ol*) and *germ vesicle* (*p*)
are very marked. So it is in regard to the organs of animals;
for the lowest of them, as I have shown, gradually *differentiate*
the *opposing* sides, the *nervous* and the *digestive*, until, as we
rise to the highest forms, we find the digestive system, or, as
the heart is a part of it, the *nutritive centre* corresponding to
the *yolk*, and the *nervous centre* corresponding to the *albumen*,
or *germinal-vesicle pole*.

When we draw a line from one of these poles to the other,
whether we do it in the lowest or the highest animals, we divide
the body exactly into *right* and *left;* that is to say, we find that
*all* animals are *double*, even man.

The brain of man is double; one half may be taken away
altogether, without affecting the mental functions, as certain
diseases have shown. The experiment of removing one half
of the brain has been tried successfully on dogs, rabbits, and
pigeons, and these animals did not lose their usual mental
powers. The paralysis of one side of a man's body, whilst the
other half retains its powers of motion, shows that the spinal
*cord* is *also double*, like the brain. The duplicity of this system

128                           BILATERALITY.

among the lower animals is much more apparent than in the higher.

Such an arrangement of the organs of the body is called *Bilaterality;* and, as you see by these diagrams, (figs. 55 to 64,) bilaterality is the *basis* upon which the animal structure is erected; and whatever modification there may be of this feature, this *type of form,* such a modification is subordinate to the *type.*

Perhaps some of you will call to mind the starfishes and sea-urchins, which you may have read about in books as being formed upon a plan which is called radiate, like the spokes of a wheel, or the divisions of an orange. It has been represented that these so-called Radiates differ essentially from all other animals, because their organs are not arranged upon the plan of bilaterality; and that whatever appearance of bilaterality — for its presence is admitted in a certain sense — there is in them, is of secondary importance. Now as the Radialists, as I may call them, have admitted, nay, have even claimed, that there is the appearance of bilaterality among Zoöphytes (Radiates), let us see what we can add to this acknowledgment. Let us refer for a moment, without intending to anticipate what I may say hereafter, to the earliest forms that have appeared on our globe; and, to make the case the more decisive, to what were probably the only representatives of their class at one time.

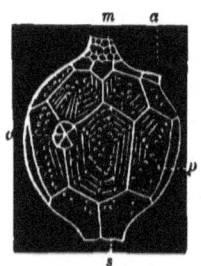

Fig. 66.

This figure (fig. 66) represents the body of one of the Crinoids, a Hemicosmites, which lived in the earliest geological age. It belongs to the same class as the trepangs, such as Caudina (fig. 54) and the starfishes (chap. x. figs. 109, 110); but, unlike them, it is attached at the end (fig. 66, s) opposite the mouth (*m*) to a stem. The principal feature is the snout-like protrusion (*m*), at the end of which is the mouth. From our knowledge of the course of the intestine

Fig. 66. *Hemicosmites.* Natural size. A fossil Encrinite, without its stem. *m,* the proboscis; *a,* the aperture to the posterior end of the intestine; *o,* the

of the living Crinoids, it has been agreed among naturalists that the small aperture, which is in the slight prominence (*a*) near the mouth, is the posterior terminus of the digestive canal. Taking, now, the two opposite extremities of this canal as the extreme points of a line, we may project that line, as a plane, through the body toward its stem, so as to divide the Crinoid into right and left portions. There is not anything about the animal which militates against this method of topography; and at the same time I would say that there is not the least rudiment of radiation in the disposition of the components of this body. There are traces, that is, scars on each side of the proboscis (*m*), where it is thought that arms were attached; but these are as definitely arranged in regard to right and left as are the arms of a cuttle-fish or squid (chap. xi. fig. 124). There are other kinds of Crinoids whose arms are much more conspicuous for their right and left arrangement than those of Hemicosmites.*

What then do you suppose would have been the decision of a naturalist had he lived at that time, — far down at the bottom of the Silurian period, — at the period of the first appearance of life upon the earth? He certainly would never have thought of such a thing as a radiate type; simply because there is nothing in these animals to suggest such an idea. I might also refer you to the embryonic or earliest stages in the growth of Zoöphytes, and you would see there also that the radiate character is either very feebly represented, or altogether absent, whilst the bilateral feature stands out prominent; but I cannot at present go into many details, as that would anticipate what you will hereafter

aperture of the reproductive organ; *p*, the plates of the shell; *s*, the point of junction of the shell with the stem. — *From Pictet.*

* For the benefit of those who may object that the ovarian opening (fig. 66, *o*) of Hemicosmites is unsymmetrically placed, and therefore is out of relation with the bilateral plane, I would propose, as an answer, to show on the same score, that the worm Bonellia (chap. xii. fig 126) is not a bilateral animal, inasmuch as the aperture of its reproductive organ is a little on one side of the median line; or that a man is not a bilateral creature, because he uses his right hand more than his left.

learn, when we come to the history of the mode of development of these animals.

But let us turn aside awhile toward the margins of these five grand divisions or groups, where the clouds, as I have said, merge into each other, and see if this idea of distinct types can be maintained.

## CHAPTER VII.

THE DISTINCTION BETWEEN ANIMALS AND PLANTS. — THE PSEUDO-INFUSORIA, THEIR PLANT-NATURE. — THE PLANT-LIKE INFUSORIA.

In the consideration of this matter, I come now to the explanation which, in a previous lecture, I promised to give you in regard to the meaning of the "*merging of the clouds*" of the *five grand divisions*. You will recollect that I stated that although some considered the five grand divisions as so many distinct groups, yet an equally large class of naturalists looked upon these divisions as so many *subdivisions* which have more or less intimate relations with each other, as if they were the *components of a vast cloud* which is divided into *five* masses, having their edges mutually merged into each other. Now I think that you will best comprehend the meaning of these relations, as to whether they are those of *consanguinity*, or *ideal*, when I have explained the relations of the classes or minor groups of each grand division. There are two distinct considerations to be held here: the one is whether these *five* grand divisions are related in the same sense as are the classes of each division, or whether they have a different relation, and what that relation is.

Now, as this distinction is based upon the very foundation of the animal kingdom, I must go back to first principles, and, as you will see presently, to the investigation of a different kind of *life characteristics* from those which were concerned in the discussion of the "*principle of life*," as I termed it in a previous lecture (p. 7). In that lecture I pointed out the distinction which exists between *organized* bodies, whether animals or plants, on the one hand, and *unorganized* bodies, mineral and chemical; on the other. But now we will take up the matter of the distinctive characters between organized life as manifested in *one* form, the *animal*, and organized life as manifested in *another form*, the *plant*.

In beginning a description of the Animal Kingdom, the first question that arises is, "*What is an animal?*" By what characters do we distinguish the *animal* from the *plant*? To the generality of people it would seem as if the question would be answered as soon as asked. No one, say they, would confound a man with a tree; a fish with a sea-weed; a coral with a mushroom, or a sponge; but to show at once how soon the very difficulty would be plunged into, by instituting such a running comparison, I will tell you that the sponge has been, for years, the centre of controversy as to its animal or vegetable nature. The common, every-day acquaintance with the sponge would not help one to distinguish it from certain kinds of corals which I could produce. There are also other species of sponges which are filled with limestone; and they cannot be distinguished from certain corals except by the closest scrutiny of the naturalist, and sometimes in the fossil state the separation is impossible. This is only one out of many instances of the kind; and that you may fully appreciate the hesitation of naturalists, of all faiths, in determining the limits between the animal and vegetable kingdoms, I will illustrate certain phenomena which occur where the uninitiated would least suspect.

From the early days of the microscope, when it was looked upon rather as a marvellous sort of plaything, up to the time, about 1826, when Ehrenberg began his researches upon the more minute organisms, all those infinitesimally small moving bodies which were seen by the earliest observers in various kinds of fluids, whether water from the ocean, or streams, ponds, stagnant pools, ditches, or decomposing fluids, such as old milk, or in sour paste, or starch, &c., were believed to be Infusorial *animals*.

Notwithstanding that Vaucher, as early as 1803, had seen that certain of these minute moving bodies burst forth from the interior of one of the common fresh-water plants (Confervæ), and developed into fixed stems and branches, like those of the plant from which they emerged; I say that in spite of this strong hint, it was not until Ehrenberg had excited the marvel of the scientific world by his disclosures of the complicated organisms of many of these moving atoms, and had stirred up a lively and sometimes rather too caustic criticism upon the correctness of his observations, and by this means had brought the microscope into use as a scientific instrument, whose tremendous power as an engine of progress, along the great road of science, I believe is but half suspected,—not until Ehrenberg had given this impulse, and there finally appeared in the field a class of observers who devoted themselves to the elucidation of the nature and relation of the so-called Infusorial animalcules,—that it was suspected that anything but animals were comprised in this group.

Finally, from the year 1843 to 1850, among other observers, Thuret was the most active in throwing doubts upon the *animality* of certain of the so-called Infusoria, which were classed together *from their similarity*, as here represented (figs. 67 to 72).

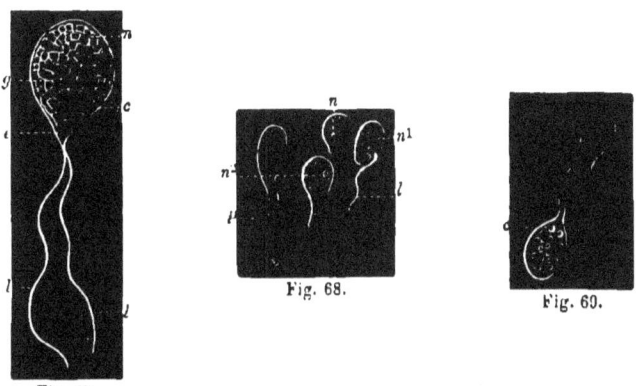

Fig. 67. *Protococcus pluvialis*. A zoöspore. 1000 diam. *c*, cell-wall; *g*, granular contents; *e*, transparent region; *n*, nucleus; *l, l*, vibrating lashes.— *Original*.

Fig. 68. *Saprolegna ferax*, in different stages of growth. 500 diam. $n^1$, body of the zoöspore; *l*, vibrating lash of the same; *n*, $n^2$, *p*, three successive degrees of development. — *Original*.

Fig. 69. *Chlamidomonas pallida*, n. sp. 500 diam. *c*, the pair of contractile vesicles. — *Original*.

134    THE DISTINCTION BETWEEN

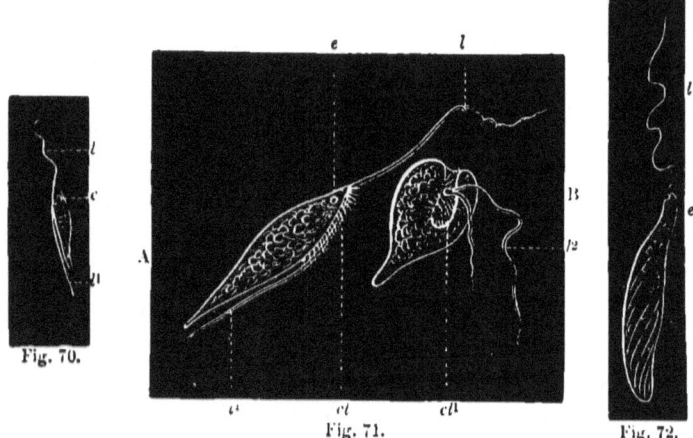

Fig. 70.          Fig. 71.          Fig. 72.

Unless I were to point out their differences, you could not distinguish the true Infusorians from the false ones. Which they are, respectively, I will explain presently.

Thuret discovered that the seeds (spores) of certain marine (Algæ) and fresh-water (Confervæ) plants have attached to them peculiar threads, or cilia, with which they move; and he also pointed out their resemblance to the published figures of certain so-called Infusorial animals.

To prove what is the real character and relations of these ciliated spores, he spent a number of years in the investigation of the mode of reproduction of a large number of different kinds of water-plants, such as are commonly called sea-weeds and pond-weeds. The process which Thuret adopted, in order to carry out his proofs to perfection, although laborious, was the only one that could be successful. At first he watched the grad-

Fig. 70. *Heteromita fusiformis.* n. sp. 500 diam. *c*, the contractile vesicle; *l*, the anterior proboci-like vibrating lash; $l^1$, the trailing lash. — *Original.*

Fig. 71. *Heteromastix proteiformis.* Nov. gen. et sp. 500 diam. *A*, an individual fully extended; *e*, the red eye-spot; *l*, the anterior lash; $l^1$, the posteriorly trailing lash; *cl*, the group of vibrating cilia. *B*, a contracted individual; $l^2$, same as *l*; $cl^1$, same as *cl*. — *Original.*

Fig. 72. *Euglena spirogyra.* Ehr. 300 diam. *e*, the eye-spot; *l*, the vibrating lash. — *Original.*

nal changes which the contents of the plants pass through in the formation of their spores, and then he doubled the proofs by tracing the growth of these self-same spores into branching plants.

Subsequent researches have confirmed the observations of Thuret, and have also carried that delicacy of distinction, which he introduced, to the highest degree of critical comparison. Moreover, it was in following out these elaborate examinations that observers have come across some of the most important physiological facts that this century has produced.

One of these facts is this: that within the boundaries of a simple circle, a series of phenomena are exhibited which constitute the components of a whole life, of *origin*, *growth*, and *reproduction*. The physiology of *life*, in this case, is reduced to, or more properly speaking, it does not rise above the simplest form of operation and manifestation.

This I think you will fully comprehend from a description of the life characters of one of the lowest organized of all plants. It is closely related to, if not identical with, the famous " Red-snow plant." In one of its states of existence it is frequently to be found not only in water, but also in damp places; and during the winter, in some localities, it gives a red color to the snow upon which it grows. The whole plant is a mere microscopic globule, so small indeed that a single one would escape the eye, but when congregated in vast numbers they make themselves visible by their color. This little globe consists of a thick, transparent, delicate shell or coat, (fig. 73, $c$,) which is filled with a reddish or brownish yellow granular mass $(g, g^1)$. The latter is most frequently found divided into two portions, $(g, g^1,)$ each one of which contains a comparatively large brownish red globule, or *nucleus* $(n, n^1)$ as it is called. In this condition the plant is known to be commencing a series of changes which result in its total metamorphosis into new individuals. The manner in which this is brought about is in this wise. When it is subjected to the action of water, the transparent envelope (which is the homologue of the stem of the thread-form aquatic plants,

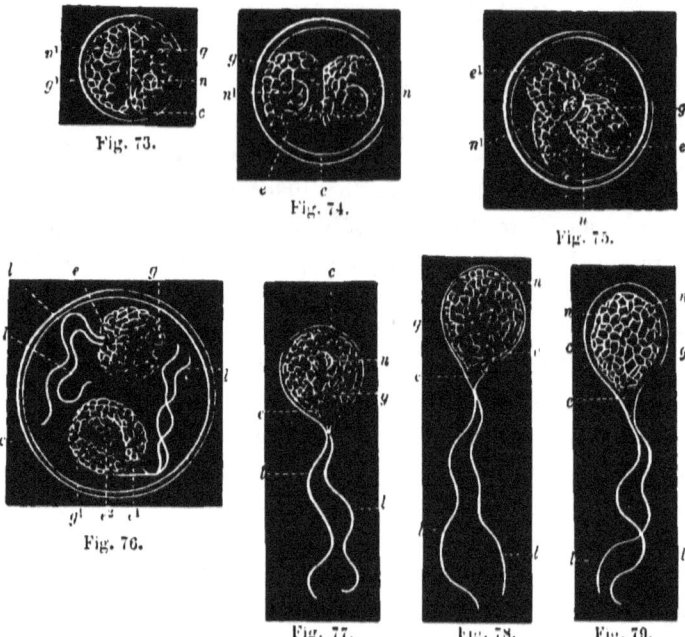

Fig. 73. Fig. 74. Fig. 75. Fig. 76. Fig. 77. Fig. 78. Fig. 79.

the Confervæ) swells to a larger size, (fig. 74, $c$,) and the two granular masses, ($g$,) increasing in transparency, become gradually tinted with a light green, whilst one end of each changes to a clear transparent area ($e$) totally devoid of granules, and the nucleus ($n$, $n^1$) enlarges considerably, and at the same time loses all of its color.* Frequently before these changes occur the two granular masses of the primary stage (fig. 73, $g$, $g^1$) divide again, each doubling itself and its nucleus, and then the changes of their color and of the nucleus, which I have just described, occur.

In this figure (fig. 75) the four masses, ($g$,) the future *spores*, have already assumed their final shape, and the narrower end

Figs. 73 to 79. *Protococcus pluvialis.* The successive phases of growth of the plant from the resting stage to maturity. $c$, the cell-wall; $g$, $g^1$, the granular cell contents; $n$, $n^1$, the nucleus; $e$, $e^1$, the transparent end of the cell; $e^2$, the transparent centre; $l$, the vibratory lashes. — *Original.*

* It is barely possible that the nucleus disappears as the transparent space develops.

($e$, $e^1$) is as pointed, transparent, and clear as in the later stages, but the nucleus ($n$, $n^1$) is as yet quite dark. This teaches us that the changes do not go on with similar steps in each individual, but yet all tend toward one end, which we find illustrated here (fig. 76). Each spore has developed, from the transparent end, ($e$, $e^1$,) a pair of thread-like bodies, ($l$, $l$,) of equal thickness throughout, which are in constant motion, writhing and lashing about as far as the increased size of the parent-cell ($c$) will allow. Plunging now with the microscopic probe within the green mass, we find its interior to be still more largely occupied by a transparent space ($c^2$) than in the first change (fig. 74, $e$) which we took note of; and moreover it is evident that this transparency is in direct continuation with the clear area at the pointed end, (fig. 76, $e$, $e^1$,) where the vibratory lashes ($l$) are attached. In this condition the spores are set free by the bursting or dissolving away of the parent-cell ($c$), and allowed to swim off through less restricted habitats than they have heretofore occupied. There is no definite aim to these movements, but each spore seems to lead a sort of indeterminate, roving life, following in an irregular line the lead of the constantly twirling double lashes.

In process of time there appears a thin, transparent film (fig. 77, $c$) on the surface of the ever-active spore; at first it is very indistinct, and the outline is indefinite, more like a halo than a sharp, light contour, but gradually it assumes greater prominence and apparent solidity, (fig. 78, $c$,) and finally it stands off from the surface of the green contents, a firm, clear, sharply defined wall, (fig. 79, $c$,) with all the characteristics of that of the parent, excepting thickness. There is at the same time another equally significant phenomenon of growth that appears during the formation of the spore-wall. Within the granular mass a faint spot (fig. 77, $n$) looms out of the darkness, and, gradually growing more intense and bright, becomes quite conspicuous; then it elongates (fig. 78, $n$) and assumes an oblong form with a constricted middle; and lastly it undergoes the process of self-division, and the two resultants (fig. 79, $n$, $n$) become the most noticeable features of the perfected spore, by their strong, brilliant, oil-like, refractive powers. This is the beginning of the end, of the completion of the cycle of development; already the preparatory step has been taken for the partition of the granular mass, by the self-division of its nucleus, and all that is required to bring it to the perfect state, the one with which we started, is a falling away of the vibratory cilia, (fig. 79, $l$,) the thickening of the cell wall, ($c$,) the self-division into two of the granular contents, ($g$,) and then the development is perfected; the offspring has become the image of the parent-plant (fig. 73).

Here we have within this little space an eternal circle of the incomings and outgoings of life; so simple in its manifestations, and yet so intensely vital; so determined toward a particular end, that one, after the contemplation of these phenomena, instinctively inquires, what chance is there now to comprehend the physiological actions and reactions of the highest and most complicated of those beings which manifest life? This question the physiological anatomist has held as a problem for the last twenty-six years, dating back to the time when the botanist Schleiden and the physiologist Schwan, in 1838, gave to the world the results of their studies upon the growth of the cells of plants and animals. Ever since that time, our ideas in regard to the high complicity of the functions of life, among the elevated classes of animals and plants, have been changing, and verging toward a more simple philosophy. In short, life, instead of being that long and often-represented entangled mesh of complications and puzzling manifestations, stands now, in the mind of the thoughtful, laboriously investigating physiologist, as a *unity*. Perhaps you can best realize this idea if I call to mind the great simplicity in the administration of medicine to the sick at the present day, as compared with the complicated operations through which the human frame was compelled to pass not many years ago. For this you may thank the student who scarcely more than a quarter of a century since bent patiently over his microscope from early morn till setting sun, watching, with almost suspended breath, the little transparent sphere which eventually became a universe of attractions for the whole circle of scientific minds.

Thuret's task in the investigation of the mode of growth and reproduction of the simplest branching water-weeds, was to show what is the nature of the apparent increase in the complicity of the progressively more elevated forms. The first and simplest step in this way is exhibited by a very common green sea-weed which is called *Bryopsis* (figs. 80, 81). Notwithstanding that it differs so much, to all appearances, from the snow-plant, and seems to be much more complicated than that, yet in reality it is scarcely more elevated in rank than the little globule whose development we have just followed through. 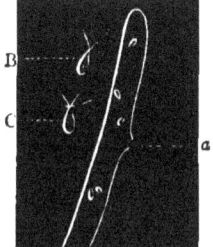 It has made an initiatory step, however, toward a higher *status* by a differentiation of its whole into stem and top, and by assuming the form of the more highly organized aquatic plants; although the organization is rendered none the more complicated by the manner in which this is done. This you will readily comprehend if I suppose, for instance, that the globular cell of the snow-plant were so plastic that you could stretch it out into a long tube, and then draw out the sides of the tube into numerous parallel, finger-like projections (like fig. 80, *a*); you would still have a single cell, with a single cavity within it, and therefore in reality no more complicated than before. Such is the condition of Bryopsis. The process of reproduction is the same as in the snow-plant; the whole contents become changed into lively, moving spores, (fig. 81, B, C,) each furnished with from two to four vibrating cilia. The entire

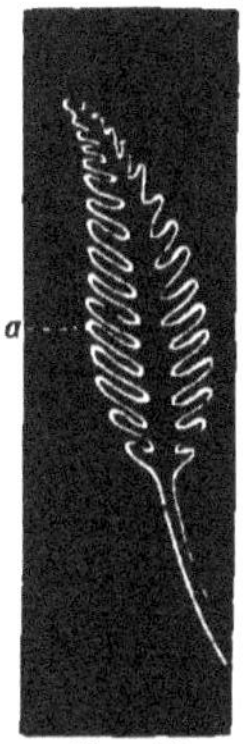

Fig. 80. *Bryopsis plumosa.* A young plant from our coast. *a*, the pinnules or lateral projections. — *Original.*

Fig. 81. *Bryopsis hypnoides.* *a*, the aperture of a pinnule like *a* of fig. 80; B, C, zoöspores, one with two, and the other with four vibratory cilia, magnified 330 diameters. — *From Thuret.*

plant does not, however, dissolve at once in order to allow the ripened spores to escape, but an aperture (fig. 81, A, *a*) is formed in the side of the finger-like projections, through which the young glides forth to a freer life. This it enjoys for a few hours, and then commences its career as a fixed plant. I will not now describe the process of transformation into this latter condition, but reserve it until we come to one of those plants in which I have myself watched the development through all its changes.

The most clearly defined step that is taken toward a higher rank, is exhibited by the partitioning of the cavity of the single plant into several chambers. This we have in the sea-weed, of which a portion is delineated here (fig. 82, A). It is called *Cladophora*. These three cavities, separated from each other by these double diaphragms, (*c*,) have each an aperture, (*a*, *a*¹,) through which the spores (B) are escaping. The whole plant, which branches considerably, is made up of similar cells, in all of which, Thuret says, spores are developed, and from which they eventually escape. The entire plant, from top to bottom, becomes a mass of seeds. It is merely a branching string of one-celled plants, each one of which is formed in the likeness of the snow-plant, Protococcus. What, in addition to its vibrating lashes, renders the zoöspore all the more like certain of the Infusoria, is a red spot in the clear space at the pointed end. This has been mistaken for and identified with the eye-spot of certain animalculæ, such as *Euglena*, (fig. 86,) &c.; but Thuret has shown it to be a mere globule of oily matter, and that, moreover, it is not always present, or is more or less indistinct.

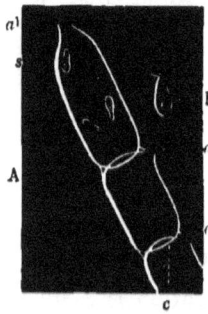

Fig. 82.

The next decided advance in rank is exhibited by those plants

Fig. 82. *Cladophora glomerata*. Ktz. A, three of the joints or cells from the end of the branch of a plant; *a*, lateral apertures; *a*¹, terminal aperture; *c*, transverse partition; *s*, young plants germinating within the cell of the parent; B, a zoöspore magnified 330 diameters. — *From Thuret.*

in which the reproductive process is confined to one part of the organism; that is, certain regions are *specialized* and devoted to a different office from that of the other merely vegetative portions, and by this specialization a *reproductive organ* is produced. Commonly, this organ is the terminal cell of the plant among the lower grades of sea-weeds and their fresh-water relatives. The one which I have represented here, (fig. 83, A,) in its natural size, grows like a white mould over dead flies and other insects which may happen to fall in the water. You may very readily raise it in a few days by throwing some flies into a jar of water, and letting it stand quiet in a warm place, when a white film of fine threads gradually makes its appearance all over the decaying body of the insect. These threads, if examined soon after they become clearly visible, will be found to be mere tubes with a sharp point at the free end, and a broad base where attached. After a while their tips lose their transparency and become whitish. If now they are examined, it will be found that this part of the plant is partitioned off from the rest, as in this figure, (fig. 83, B,) and the contents are little yellowish, globular bodies crowded together as close as they can lie. Presently this whole mass begins to be agitated, and the globular bodies tremble from an apparently invisible cause, reminding one of a commu-

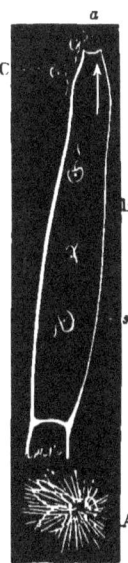

Fig. 83.

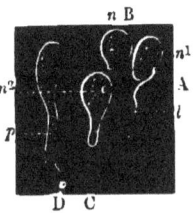

Fig. 84.

Fig. 83. *Saprolegnia ferax?* A, a group of plants growing on a dead fly; B, the tip of a plant magnified 250 diameters; *a*, its terminal aperture; *s*, zoöspores; C, zoöspores just escaped from the plant. — *Original.*

Fig. 84. *Saprolegnia ferax?* Same as fig. 83. Various stages of growth after the escape of the zoöspores; A, a ripe zoöspore; $n^1$, its body; *l*, its vibratory lash. B, the first stage of the plant-growth; *n*, its nucleus. C, second stage; $n^2$, its nucleus. D, the young plant just attaching itself; *p*, the basal end. 500 diam. — *Original.*

nity of bees when swarming; then the cell which confines them seems to be convulsed, and swells and stretches now and then, until finally the end (*a*) bursts open, and the swarming globules (C) are ejected in a body by the contracting cell. Occasionally a few are left behind, (*s*,) and commence to develop within the parent-cell. I wish particularly to draw your attention to the form of the spores (fig. 83, C, and fig. 84, A) of *Saprolegna*, on account of their resemblance to certain Infusoria. Compare the zoöspore, (fig. 84, A,) with its eye-spot-like nucleus, and its single* vibrating lash (*l*) attached at the bottom of a notch at one side, with the Euglena, (fig. 86,) and you will not wonder that the spores of sea-weeds have been mistaken for animalcules. It was no difficult matter in this case to prove that they were the genuine seeds of the plant from which they came, and not its parasites; for in an hour and a quarter after they were set free, and during which time I watched them constantly, they began to stop their roving, and settling down upon *terra firma* with a sort of sideway motion, as if trying to wedge themselves into some hollow, they became globular, and insensibly lost the vibratile cilium by what appeared to be a process of deliquescence. In the course of two or three hours after this, one side of the globule began to show distinct signs of growth by a slight protuberance, (fig. 84, B,) and consentaneously a distinct trace of a cell-wall appeared upon its surface, in the same way as we have seen it develop in the snow-plant (p. 137). In time the protuberance lengthens so as to give the young plant a pear-shaped figure, (C,) and then, continuing to elongate, it becomes tubular in form, (D,) and, the rounded end at the same time growing comparatively narrower, the whole soon assumes the shape of the parent stock. It is not necessary to watch the growth of a spore through all its phases up to the full-grown state, as every step beyond what I have depicted here can be found exemplified

---

* As these spores have not the *two* terminally attached cilia which Thuret figures, it would seem that this plant must be not only specifically but generically distinct from *Saprolegna ferax*.

in a group of plants, upon placing the whole colony under the microscope.

We have already had before our eyes two methods by which plants rise from a lower to a higher rank, namely, by the formation of distinct cells in a continued series, and by the specialization of parts for a particular function. The complication is still further increased by the development of cells in a lateral direction, so as to form a leaf-like plant. The common, light green, parchmenty sea-weed, *Ulva*, everywhere upon our rocky coast known under the name of green *Dulse*, or *Laver*, is an example of this kind; and the olive-colored, or brown, leather-apron-like sea-weed, *Laminaria*, exemplifies the increase of growth of cells, not only longitudinally and transversely, but at right angles to the latter direction. In this whip-cord-like plant, (fig. 85, A,) known in science as *Chorda*, which is very common in rocky bays, the lateral growth of cells is such as to produce a round figure, from slender base to tapering tip. The spores (B, C) resemble those of Saprolegna, but have a red eye-spot. According to Thuret's observations, they are developed in the superficial cells of the cord This, I think, will suffice to illustrate the tendencies of the march of development from the lower toward the higher forms of plant-life, as contrasted with a similar procedure among the inferior grades of animal life, which I have described to you in a former lecture (pages 9 to 14). But let us now turn again more particularly to the supreme result of these investigations, which was to show that certain kinds of so-called Infusoria were not animals but plants.

Fig. 85.

Here, then, was a new set of phenomena to be investigated by the physiologist. Zoölogists said at once that the fact that a body moved from place to place was no longer a criterion of its animality; nor could they fall back upon another fact, which

Fig. 85. *Chorda filum*. A, a plant; B, C, zoöspores, magnified 330 diameters. — B, C, *from Thuret*.

was so long a strong basis, and apparently invincible, namely, that vibratile cilia are indubitable indications of an animal nature in the body which possesses them, for we have seen how active those cilia are which are attached to the spores of the lowly organized plants. They also said that the red "*eye-spot*" — as Ehrenberg believed it to be, and consequently deemed it a nerve of sense, — had also lost character, and was reduced to a mere globule of oily matter which was sometimes present in, and sometimes absent from, the rapidly changing contents of the flitting, spasmodic spores. It is true, however, that certain undoubted animals, Euglena (fig. 86) for instance, also have this red, eye-like spot; but its presence in spores of plants deprives it of all character as a mark of distinction unless accompanied by other diagnostic features.

Casting about from point to point, endeavoring to find some safe foundation upon which to raise a firmer and more lasting framework, naturalists came to the conclusion to settle down upon *contractility* as one reliable character, and the *absorption of food* as another, and probably the more trustworthy; but both were far beyond suspicion, as was then thought.

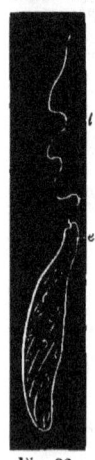

Fig. 86.

Let us see, now, what proofs of animality we may derive from some of the infusoria which most resemble the sea-weed-spores. I will take, for example, one of the most familiar and common of the animalcules; it is known under the name of *Euglena*, or the *Eye-animalcule*, (fig. 86). The resemblance in character to the spore of Saprolegna is heightened by the fact that the vibratile cilium (fig. 86, *l*) of the Euglena, like that of the former, emerges from a notch near the narrower end, and moreover there is close to it a nucleus-like spot; but the latter in Euglena is red, and is known as the red eye-spot (*e*). The spores of Cladophora (fig. 82) and Chorda (fig. 85), you will recollect, also have a red spot near the vibratile lashes;

Fig. 86. *Euglena spirogyra.* Ehr. 300 diam. The Eye-animalcule. *e*, the red eye-spot; *l*, the vibratory lash. — *Original.*

so that the matter of color does not affect anything in this comparison. So far, then, the argument holds good for their close relationship; but when I assure you that the Euglena as figured here is only one of the many forms which the same individual may assume from moment to moment, and that you may see it change before your eyes, almost as quick as thought, from this elongate figure (fig. 86) to one like this pear-shaped infusorian (fig. 88, B), and that in the next second it is stretched out and pointed at each end like a spindle, as in this figure, (fig. 88, A,) and, in the midst of these various elongations and contractions, exhibits an extreme degree of flexibility, at times fairly doubling itself up end to end, as you would fold a strip of india-rubber, you cannot fail to appreciate the marked diversity of character between the two objects in question. How far this distinction goes, we will not stop to discuss until we have seen these other kinds of infusorians. The separation apparently grows wider yet when we learn that Euglena takes in food in the form of solid particles, and stores them away in globular cavities, sometimes called the digestive vacuoles.

There is still another peculiarity among these spore-like Infusoria which does not find its parallel in the sea-weeds. I refer to the contractile vesicle. I was fortunate in obtaining another infusorian, *Chlamidomonas*, (fig. 87,) which resembled the spores of some of the seaweeds so closely, not only in form and actions, but also in size, which in the doubly ciliated spores is very minute, that, were it not for one almost inconspicuous character, I should certainly have taken it for the colorless spore of some water-weed, or of something like the snow-plant. That character was exemplified in its double contractile vesicle (*c*). These vesicles appeared and disappeared alternately, or sometimes both together, with considerable rapidity, and yet, notwithstanding

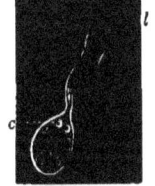

Fig. 87.

Fig. 87. *Chlamidomonas pallida*, n. sp. 500 diam. *c*, the pair of contractile vesicles; *l*, the vibratory lashes. — *Original.*

their minuteness, when once recognized, it was an easy matter to see that they had the physiognomy and habit, as I might say, of the contractile vesicles of well-known infusorians.

We have now noted three diagnostic features in the Infusoria which we did not observe in the plant-spores; but let us go on still further with the animals for the purpose of getting a broader basis of comparison; one that will aid us in discovering the nature of the differences between the developmental process, from the lower to the higher ranks, of the Infusoria, and the corresponding process in the sea-weeds. Here is an infuso-

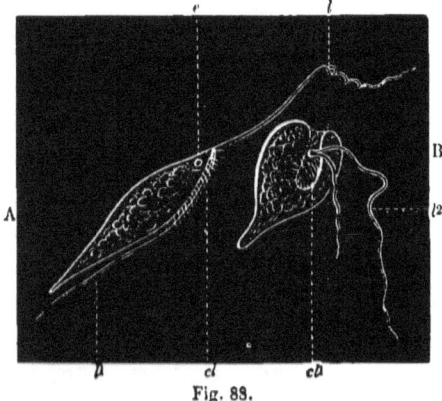

Fig. 88.

rian (fig. 88) from fresh water, which, although it has a pretty strong resemblance to Euglena, heightened by the presence of a red eye-spot, ($e$,) will be found, upon investigation, to possess some additional and decidedly different characters. In the first place, it has two vibrating lashes, ($l$, $l^1$,) which differ remarkably among themselves both in position and character. One of them is always carried in front like a sort of proboscis, ($l$,) and in fact it seems to have the office of such an organ, like that of the elephant, to feel and to take hold of objects. I must confess that I was struck with astonishment at the apparent intelligence with which the infusorian extended and twisted and turned and felt about with this extraordinarily muscular organ. Never did an elephant seem to use his trunk with more thoughtfulness. With like control did the animal also use the other lash, ($l^1$,) always keeping it turned

Fig. 88. *Heteromastix proteiformis*, nov. gen. et sp. 500 diam. A, an individual fully extended; $e$, the red eye-spot; $l$, the anterior lash; $l^1$, the posteriorly trailing lash; $cl$, the group of vibrating cilia; B, an under-side view of a contracted individual; $l^2$, the same as $l$; $cl^1$, the same as $cl$. — *Original.*

back along its body; so that it formed a kind of movable keel, when the little creature glided through its watery element, or was used to sway it from side to side, or oftentimes to raise it up on its tail by forming a prop, as we see it in this other figure (fig. 88, B). The motory or propelling power, on the other hand, is restricted, at least in the greatest measure, to another kind of vibratile cilia. These are very short, and are crowded together in great numbers in a broad furrow or depression, ($cl$, $cl^1$,) which extends over half the length of the body, along its inferior, middle line. When the body is turned over, and the anterior end retracted and swelled out sideways, the furrow becomes quite conspicuous, (fig. 88, B,) and the extent of the group of minor cilia is easily ascertained. They are very minute and in constant motion, propelling the body backwards and forwards, up and down, to the right or left, according as it is steered by the trailing lash which extends along its length. Thus it is, that, although similar in form, a diversity of functions is laid upon these three kinds of cilia that amounts to the most marked specialization, through the simplest means; in fact, so simple that the eye cannot detect them in any form beside that of proportion and position, and certainly not in the intimate structure of these bodies. The whole body, too, possesses a flexibility and extensibility scarcely inferior to its cilia; at one moment it is darting through the water, sharp as a lance at both ends, and at the next it is as round as a ball, or worming its way through tortuous passages with every possible degree of flexure short of actually tying itself into a knot.

When, now, I turn your attention to one of the next succeeding higher forms, this one here, (fig. 89,) for instance, *Ceratium*, some of you may perhaps recognize in the one which we have just studied a transition from those, like Chlamidomonas (fig. 87) and Euglena (fig. 86), which have but one kind of vibratory lash to those which possess two forms of such organs specialized to the highest degree. In this new subject the smaller cilia (fig. 89, $w$, $w^1$) are methodically disposed in a linear series along the edge of the circumambient annular furrow which

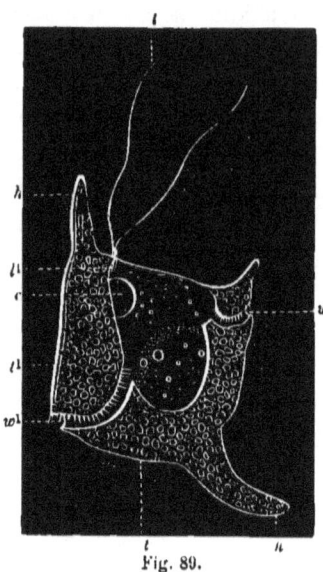

Fig. 89.

separates the two parts of the minutely sculptured shell ($l$, $l^1$). It is in this animal that, as we ascend from the lower to the higher kinds of Infusoria, we find for the first time a disposition of the cilia which prevails so generally among the aristocrats of their class. I need but to remind you of the Stentor, which I described when speaking of the self-division of Infusoria (p. 62, and fig. 30) and other animals. The habits of the Ceratium, whilst swimming, correspond to the arrangement of these cilia: it progresses with a spinning motion, boring its way through the fluid by being whirled, like a wheel on its axis, by means of the transverse vibrating motion of the belt of cilia, whilst its double proboscis ($l$, $l^1$) plays the part of a tactile organ.

I will detain you here with but one more example of the progressive series, simply that the investigation shall have fairly entered the bounds which include the true, so-called *ciliated Infusoria*. You would hardly suspect at first glance that this oval

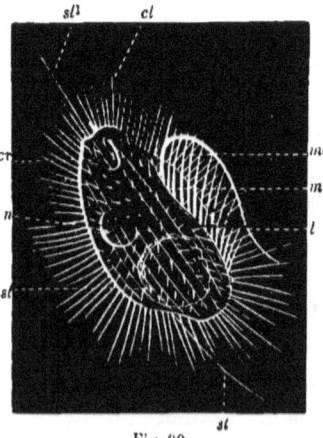

Fig. 90.

figure here (fig. 90) possessed any configurative relationship to the trumpet-shaped Stentor (fig. 30); but yet, the young (fig. 91) of the latter, when placed beside this, *Pleuronema*, cannot fail to strike you with its resemblance to it (compare figs. 90 and 91). I hope hereafter to prove to you conclusively that the similarity is based upon an idea of form which is common to both of them, and, in fact, to all Protozoa.

What I wish now to show in the Pleuronema is the triple, or I might say even the quadruple, diversity of the vibrating cilia, or in other words, a quadruple specialization of one type of organs, by their manifold offices ranking their possessors above those of their class which attain to a less degree of complicity in this respect. The most prominent of these cilia are those which are arranged in longitudinal rows (fig. 90, cl) over nearly the whole extent of the body, and which most frequently are seen in a quiet state, projecting far out from the surface like so many fine, rigid bristles. In fact, the motions of this animal are so lightning-like in rapidity that I have never seen this form of cilia except when the body is in a quiet state, and therefore I judge that, as they do not move then, they are the principal organs of locomotion. There is on the right side a group of much more heavily built cilia, (mc,) which project from the oblique furrow in which the mouth (m) is set. They are more particularly devoted to producing currents in which the particles of food may be brought to the mouth. We see, also, projecting from the forward end of the oblique furrow, and near the anterior edge of the mouth, (m,) one of those proboscis-like lashes (l) which are so characteristic of the lower, ciliate Infusoria; but yet it would not seem to have the same office as in the latter, since it is usually held in this position, apparently as rigid as if it were a wire; and only now and then does it move, by a sudden

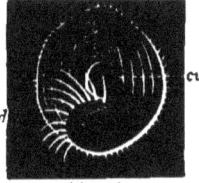

Fig. 91.

Fig. 89. *Ceratium cornutum.* Clap. 300 diam. A marine infusorian, covered by a reticulated shell (testa). *t*, *t¹*, the two halves of the testa; *h*, the horns of *t*; *l*, the vibratory lashes; *l¹*, base of *l*; *w*, *w¹*, transverse rows of vibrating cilia; *c*, the contractile [?]vesicle. — *From Claparède.*

Fig. 90. *Pleuronema instabilis*, n. sp. 1000 diam. From fresh water. *m*, the mouth; *st*, the food gathered in one mass, *mc*, large cilia in the vestibule of the mouth; *l*, the single vibrating lash projecting from *m*; *cl*, vibratory cilia covering the body in rows; *sl*, the posterior, and *sl¹*, the anterior saltatory cilium; *cv*, the contractile vesicle; *n*, the reproductive organ.—*Original.*

Fig. 91. *Stentor polymorphus*, Ehr. 300 diam. A very young individual, turned upside down to compare with fig. 90. *d*, edge of the oblique furrow; *cv*, the contractile vesicle. — *From Claparède.*

jerk, and disappears in the oblique furrow; probably acting there in concert with the other cilia in the introduction of food into the mouth. The fourth and last kind of cilia which I have to speak of are two excessively faint, very long, and quite large, bristle-like filaments ($sl$, $sl^1$) which project from each end of the body. The straight one ($sl^1$) always precedes when the creature is in motion, and the curved one ($sl$) is attached a little to the left of the posterior end of the body. Both are always rigid when the animal is not in motion.— but yet there can be no doubt that they are flexible, for at times they disappear suddenly, and probably are bent under the body. What their office is I cannot say, but conjecture, from their resemblance to what are called the saltatory bristles of other infusorians, that they are used as accessory means of sudden propulsion, or leaping, — a habit which seems to be the most frequent mode of leaving any point at which the creature has fairly come to a stand-still. The contractile vesicle ($cv$) lies close to the forward end of the body, and corresponds in activity to the vivacity of the motions of the latter. It contracts every ten seconds, and with more vigor than any other that I know of. It is very conspicuous, as it is two thirds of the time in an expanded state; and disappears and reappears like the sudden closing and opening of a large eye. I have already indicated the position of the mouth as being near the broader anterior end of the oblique furrow, but again speak of it here in order to make the description of the digestive system complete. From the mouth ($m$) the food passes directly into the general cavity without going through any throat, and most frequently combines in large masses ($st$). The presence of a reproductive organ, which we find here ($n$) in the form of a clear, colorless, globular body, when added to all the other systems which I have mentioned, puts this animal in the condition of a fully organized, ciliated infusorian; and would seem to give us full warrant for believing it to be the culmination of a progressive development, whose tendency is to pass through such forms of animate organization as we have just been tracing in the successively more and more complicated creatures whose images are before us.

But now, to go back to the lowest of them, the spore-like forms, let me recall, in a word, what are the animal-like characteristics of their organization. They are, *contractility* of the body, the *pulsation* of an organ called the *contractile vesicle*, and the *introception of food*,—none of which have been recognized in the spores of any of the aquatic plants which most resemble these infusorians in external conformation.

Now this would appear to settle the confusion which had mingled such a mass of heterogeneous material; and the zoölogist, on the one hand, would be enabled to draw his inferences without fear of imposing upon animals those characteristics which, on the other hand, the botanist might claim for the special objects of his study. But it would seem to be a vain hope, and the work of the physiologist and the microscopist has but just begun; for to this day there remains a doubt as to the animal or vegetable nature of certain forms, which have characters that lead on the one side to *plants*, and on the other to *animals*.

The investigation of this apparently twofold relation is perhaps by far the most important among all the researches of the present day upon the subject of spontaneous generation and the development theory; for *if* this twofold relation is not apparent, but *real*, then there is no distinction between animals and plants.

What test, therefore, can we apply to those doubtful forms which I have spoken of, in order to prove that they are either living, sentient beings, or that they are as devoid of sensation as the grass of the field. It would seem almost absurd to undertake such a thing as the proving of a self-evident difference between a weed and the animal which feeds upon it; but when we reflect upon how many characters the two have in common, such as life, growth, circulation of fluids through the tissues, a remarkably similar, in fact, almost identical cellular structure in many cases, and finally a closely related mode of reproduction;—I say in reflecting upon these more or less common characteristics, the apparent absurdity rises to a dignity worthy of the mind of the most gifted genius; and when naturalists are accused, by the unthinking, of trifling over these little things, these minute drops of jelly, these moving points, these strings of granules like a row of beads, let us reply, that a greater than we allows " not even a sparrow to fall to the ground without His notice." Let proud man know that he was once as simple and senseless as that little drop of gum-like fluid, the incipient egg, which is far more lowly organized than these " *doubtful forms.*"

## CHAPTER VIII.

THE PHYTOZOA, OR PLANT-ANIMALS. — THEIR RELATION TO UNDOUBTED ANIMALS.

But to the test now. One of the most perplexing, and yet exceedingly interesting, of all these doubtful forms, is the so-

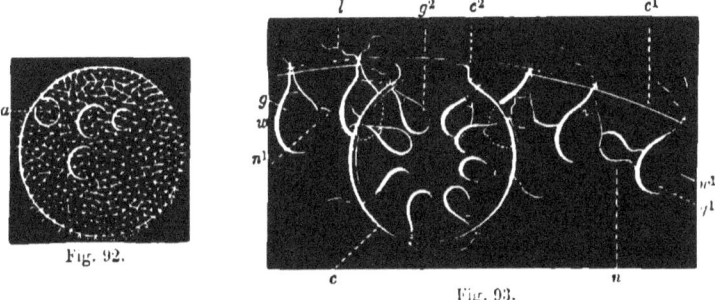

Fig. 92.

Fig. 93.

called *Globe-animalcule*," (figs. 92, 93,) or *Volvox*, as it is named, on account of its constant revolving motion whilst it progresses through the water. It is a minute, green, globular body, just visible to the naked eye, everywhere in our ponds and ditches, so common among the water-weeds that every dip of the gauze fishing-net collects them in abundance. When observed with a low magnifying power, it resembles a globular net-work (fig. 92) with green spots set in, like emeralds, at the junction of the threads.

Fig. 92. *Volvox globator.* Ehr. 50 diam. The " Globe-animalcule." A specimen containing four young, of which one ($a$) is seen in profile attached to the inner face of the hollow globe. — *Original*.

Fig. 93. *Volvox globator.* Ehr. 600 diam. A portion of the gelatinous envelope of a full-grown Volvox, with a very young one attached to its inner face. $c^1$, the envelope of the old stock; $c$, the envelope of the young; $c^2$, point of attachment of $c$ to $c^1$; $g$, $g^1$, the zoöspores of the parent stock; $l$, vibratory lashes projecting through $c^1$; $g^2$, zoöspores of the young with their vibratory lashes projecting through $c$; $w$, $w^1$, cell-wall of $g$, $g^1$; $n$, $n^1$, threads attaching the zoöspores to each other. — *Original*.

If, now, we magnify it very highly, and get a profile view of it, we shall see that it is a hollow sphere whose crust is a thin transparent gelatinous substance, (fig. 93, $c^1$,) of equal depth throughout, and that it very closely resembles the cell-wall of the snow-plant (fig. 76, $c$). This it does, too, in other respects, as we shall see presently. The emerald spots which I spoke of just now turn out to be a colony of oval bodies (fig. 93, $g, g^1$) almost identical in character with the spores of the snow-plant, (figs. 76 to 79, $g$,) but with this difference, that, instead of being loose in the cavity of the cell, their points are all fixed in the thickness of the gelatinous shell, (fig. 93, $c^1$,) and the double lashes ($l$) vibrate on its outside. By the combined action of these vibratory cilia, the sphere is kept in motion, revolving from place to place with a never-ceasing meaningless aim. The threads ($n, n^1$) of the network connect the spores ($g, g^1$) with one another by attaching themselves to the sides of the latter, and in fact they seem to be portions of the substance, or cell-wall, of these spores drawn out into fine threads.

There is also additional evidence of the relation of Volvox to the snow-plant in the mode of reproduction. As in the latter the zoöspore of Volvox is covered by a gelatinous layer ($w, w^1$) which corresponds to a cell-wall. Now, when new individuals are formed, the first step taken is a self-division of the green contents of a zoöspore into two masses. This being done, the once single zoöspore is in the condition of the like body of the snow-plant after undergoing a similar process, (fig. 74,) and in fact it is not always possible to distinguish the one from the other at this stage. When the division has occurred again, it resembles the cell of the snow-plant with four young zoöspores, (fig. 75,) even to the position of the pointed ends ($e, e^1$) which lie next the cell-wall. And so the self-division goes on until a large number of zoöspores are produced, (fig. 93, $g^2$,) whilst the original wall of the primarily dividing zoöspore enlarges (fig. 93, $c$) to accommodate the increasing colony, at the same time adhering to the wall of the parent stock ($c^1$) by a sort of neck ($c^2$) until a certain period, when the young Volvox escapes by breaking through the shell of its primogenitor, and swims away by the help of the cilia of its own zoöspores.

Thus it would appear that not only by its structural relations, but also by its developmental process, Volvox stands in close affinity to the snow-plant; but there are certain other beings which lend still more powerful evidence as to its plant-nature. There is a so-called fresh-water plant, *Gonium*, which differs from Volvox only in being flat instead of globular; just as if a Volvox were cut into halves, and each hemisphere flattened out into a leaf-like body. The Gonium usually grows in minute square patches, and has but a few zoöspores; but there are others of its aquatic relatives, such as *Tetrasporæ*, which develop to a larger size, and resemble the faded leaves of curled lettuce. From these the transition seems indubitable to other leaf-like Algæ, but which do not possess ciliated zoöspores except at stated periods. Such are Monostroma, whose cells are in a single pavement-like layer, or Ulva, already mentioned in another place, (p. 143,) whose cells are in two layers, or Laminaria, Chorda, &c., as I have already indicated (p. 143) when speaking of Thuret's investigations.

All this would appear to be perfectly clear, were it not for one fact which I have not yet mentioned, and that is to this effect. Certain of these so-called Algæ possess an organ which hitherto has been looked upon as exclusively an animal property, — I mean a *contractile vesicle*. Not only does Volvox possess this organ, but the plant-like Gonium has it in every one of its zoöspores. What, then, is there to distinguish the Gonium, when in its young zoöspore state, from the Chlamidomonas, (p. 145, fig. 87,) which I described in another place, or still less from the Euglena, (p. 144, fig. 86,) with its green contents, or Heteromastix, Ceratium, Pleuronema, &c., which I have shown to be closely related to each other (pp. 144 to 150) in an ascending series from the zoöspore-like Chlamidomonas to the highly organized Pleuronema? Indeed, some of the most recent investigators, like Messrs. Claparède and Lachman, have unhesitatingly concluded that Volvox and Gonium are animals, simply on the grounds of their having this contractile vesicle.

There is one feature, however, which would yet seem to militate against their being animals, and that is, they have never been seen to take in food; but as a set-off to this, there are beings of an undoubted animal nature which do not take in food into a stomach. Some of the intestinal worms, Tænia, have no stomach, but absorb their fluid nourishment through the porous skin of the body; and therefore we are allowed to suppose that Volvox may possibly do the same.

So you see that although the Volvox, Gonium, and I will add many other related kinds of animalcules, have one or two strong *animal characteristics*, the most positive of which is the *contractile vesicle*, yet, on the other hand, their *vegetable* characteristics, especially in the mode of reproduction, are just as strongly in favor of the botanist's claim to them. On account of these mutually common characteristics, the lowest forms have been called, by some naturalists, *Phytozoa*, i. e., *plant-animals*.

Commencing now with these "Plant-animals," let us see where they will lead us if we follow them through their successively rising grades of complication. As I have already made you acquainted with the characters of all those which I shall mention in this series, I need only to call them to mind in the same succession as that in which I formerly described them, in order that you may trace their progress from the lower to the higher types. First among them comes the doubtful Volvox, (p. 153, fig. 92,) then the more decidedly animal organization of Euglena, (p. 144, fig. 86,) and then Chlamidomonas, (p. 145, fig. 87,) Heteromastix, (p. 146, fig. 88,) Ceratium, (p. 148, fig. 89,) Pleuronema, (p. 148, fig. 90,) &c., &c., until, if we go on, the highest of the Protozoa crown the series. Or I might commence with another of the doubtful plant-animals, the Sponge, (p. 41, fig. 21,) and through the Actinophrys, (p. 44, fig. 22,) pass to Lithocampe, (p. 49, fig. 23,) Zoöteira, (p. 50, fig. 24,) and Podophrya, (p. 51, fig. 25,) and so on to the highest of the Protozoa.

Having at last, then, passed the boundaries between vegetable and animal life, and fairly stepped within the circle of *sentient* beings, the next step that we would most naturally take is to ascertain what are the relations which exist among them.

## CHAPTER IX.

#### THE SYMBOLICAL ANIMAL. — THE PROTOZOA.

You will recollect that in presenting to you the diagrams (pp. 122, 123) of the five types of animals, in which was represented the *progressive idea* of the development of the nervous system, I kept that idea in view as the prominent feature; and this I did because I had a particular aim, which was simply to show the Animal Kingdom as a *whole*, as it were *one animal, the typical or symbolical animal, in its development from the lowest expression of life to its highest manifestation.*

The older embryologists held it as a theory that the highest animals actually do pass through all the phases of the lower life, from the Monad, the jelly-like moving sphere, up through the long scale of the intermediate grades, to the adult stage of the animals in question. Thus, they said, a quadruped, *e. g.*, a Rabbit, commences as an Infusorian, then it passes into a Polyp, then into a Molluscan, then an Articulate, worm-like creature, then a Fish, then a Reptile, then a Bird, and finally takes on its own particular form. Now, what misled embryologists into this error is this undeniable and well-established fact, namely, the rabbit begins as a fish-like creature in structure, then in the process of growth it assumes a reptile-like, and then a bird-like, state of organization, and finally it becomes fully developed. All this being so clear to their eyes, they were led to make certain other deductions, and then to assume that it commences with the lowest forms, such as I have just pointed out.

No doubt in one sense there is truth in these latter assumptions, but in another sense they are untrue; it is true in this way, that as an *animal*, the rabbit goes through certain phases which are common to all animals; for example, it begins as *an egg*; it has a right and a left, *i. e.*, it is *bilateral*; it has a *nervous system* and a *digestive system*, both in the lowest possible form, and at the beginning no more definitely fixed than in the Infusorian. Then in the second place these organs begin to have definiteness, but yet of that loose kind which we find in Polyps and Hydras; the digestive and circulatory systems are not separated; so it is in Polyps, Corals, and jelly-fishes; but developing further, the digestive and circulatory systems appear partially distinct, as in Echini and starfishes; then more so, as in shell-fish, (Mollusca,) and at the same time the nervous system becomes prominent in a great ganglion, the *brain*, (curiously enough, too, resembling the nervous system of the highest Mollusca in proportions,) which extends backwards, sending out as it were, toward the tail, a ribbon-like projection along the middle line of the body, as in the highest kinds of Insecta.

But here the relation stops, and here it is that the older embryologists mistook their way; for, notwithstanding these parallelisms of growth which I have traced out, they have altogether different relations to each other in the rabbit from what obtain in the infusorian, or polyp, or starfish, or shell-fish, or insect.

This difference in the *relation* of parts is the *dividing line*. As I told you in another place, in considering the *typical forms* of life, it is the *relation* and not the *nature* of a substance which is to be taken into account. *Relation should be the ruling standard.*

In the rabbit, notwithstanding the trend or course which it pursues along the animal-base-line, the *relations* into which its gradually developing organs are brought are different from those in which the organs of an infusorian, or polyp, or molluscan, or insect are arranged when developing; and all four of these last are as different in this respect from each other as they are from the vertebrate rabbit.

But in the ideal diagrams of the types you do not see these relations exemplified, so that you could detect them. In fact, I might say that I have in a measure concealed them by the position in which the figures are placed, and by giving prominency to the main idea, *bilaterality*.

But I did this, as I have just said, that I might represent a progressive idea of the development of the nervous system, and the gradual differentiation of the digestive and circulatory systems, throughout all animals; making the animal kingdom *one symbolical animal*, passing in its development from the lowest to the highest expressions of life.

Thus in the Protozoan stage, (see diagrams, page 122,) which is the lowest, the nervous system, although formless and scattered, holds a certain position in reference to the other organs. In the *Zoöphyte* stage the nervous system is more collected, and arranged in reference to *right* and *left*, and above and below. In the *Molluscan* phase this system becomes more concentrated toward the side opposite to that on which the heart is placed. In the *Articulate* state the concentration is carried out still further; and finally, in the *Vertebrate* condition the nervous system attains its highest confluence toward the median line, and in its tendency toward the head; as had already been foreshadowed in the highest Mollusca, *i. e.*, cuttle-fishes, and in Insecta.

And now let us turn from these general relations to a consideration of those special kinds which characterize each grand division by itself. In order to make these special *type relations* clear to your eyes, I must use a different set of illustrations, as I do not wish to superimpose on these ideal figures of the progress of the typical animal-development anything which may obscure the *main idea*. It is best that it should stand alone for the purpose of further illustration when I come to the discussion of the typical idea of each separate grand division.

I will commence with the lowest type of undoubted animals, the *Protozoa*, and proceed upward, so as to show how the two ideas, e. g., *bilaterality*, and the *type of division*, are related, and how the *type* of *each grand* division is superimposed upon the *symbolic, bilateral idea.*

*Protozoa.* The type of this division is found in its relation to a *spiral;* it is the oblique or *spiral type.* If you will call to mind now, by the help of these figures of Stentor, (figs. 30 to 33, pp. 62, 63,) what I have told you about its structure, I think you will readily understand me when I speak of the oblique or spiral type of the Protozoa; yet there are others than Stentor in which this character is more evident; and that you may see how extended and universal it is, I will place before you several kinds of Infusorians with as great a diversity in this respect as it is possible to imagine; and at the same time will show you that the various organs in all of these have a common physiognomy, such as might be expected if they belong to the same type. As you are already familiar with one of the higher forms of this group, I mean Stentor, you will the more readily, than at any other time, comprehend the relations of the organs of the highest of them all, if I speak of it first.

The animal in question is one of the Vorticellæ, or bell-shaped animalcules, technically called *Epistylis*. This diagram (fig. 94) represents a colony, or, I might say, a tree of these little bells tipping the ends of the forked branches. You may find it, along with a multitude of its nearest relatives, clothing the loose roots of trees and shrubs, which project into the fresh-water ditches and pools, as if with a thick mould; and may recognize it, and distinguish it from other kinds, with the naked eye, by its not contracting and shrinking into a smaller mass when touched. This is owing to the uncontractile, rigid character of the stem and branches (fig. 95, *p*). The body is, however, very mobile, at one time expanding to a broad, slightly oblique, bell-shaped figure, and at another, rolling in the edge (*d*) of the bell, and assuming a close globular shape, hardly recognizable as the active creature of a moment before. This contractility re-

Fig. 94.

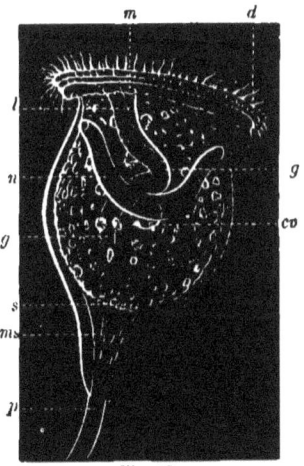

Fig. 95.

Fig. 94. *Epistylis flavicans*. Ehr. A single, many-forked colony of Bell-animalcules. Slightly magnified. — *Original*.

Fig. 95. One of the bell-shaped bodies of fig. 94, magnified 250 diameters.

sides in a thick layer of muscular substance which envelops the whole body. At the tip of the bell the muscle has the form of a conical bundle of fibres (*ms*) which, at the narrower end, is attached to the top of the stem (*p*), and at the other to the bottom (*s*) of the digestive cavity. The bell appears at the first glance to have scarcely any obliquity of form, but that is because its edge (*d*) is so prominent a feature with its thick folds and triple row of numerous vibratile cilia, that it hides, as it were, the true character of the *disc* which it surrounds. The position of its broad throat will, however, fully vindicate its claim to a relationship with the spiral type of conformation. The mouth (*m*) opens close to the edge, and leads directly into the funnel-shaped throat, or œsophagus (*g*). The latter plunges obliquely toward the left side of the body, and at the same time downwards and inwardly with a gentle spiral turn, and then diving still deeper it coils around to its right ($g^1$), and thins out almost to a point near the bottom of the digestive cavity. The food is passed down this inclined funnel into the digestive cavity by the aid of vibratile cilia which line it from mouth to bottom. What is the proper use of this single long vibrating lash (*l*), which projects outwardly from so great a depth, I cannot say precisely; but it appears to be very active in repelling certain particles of unwelcome matter, apparently not palatable to the animal. That the digestive cavity is one vast hollow, would seem to be proved by the fact, that, as I have frequently seen, the whole mass of food which pervades it everywhere, from the edge of the bell to the top of the conical muscle, (*ms*,) revolves like a great transparent ball of compact jelly and coarse, imbedded granules, one side of which passes up on the right and then down on the left with a slow and measured turn; and with such a uniformity of motion that any single granule may be selected and watched throughout a whole revolution, without the least doubt as to its course in the body. You

---

*p*, the stem; *d*, the flat spiral of vibratory cilia at the edge of the disc; *ms*, the muscle; *m* to *s*, the depth of the digestive cavity; *m*, the mouth; *g*, $g^1$, the throat; *l*, the single vibratory lash which projects from the depths of the throat; *cv*, the contractile vesicle; *n*, the reproductive organ. — *Original*.

will see, however, that this revolving matter must be a fluid, in order to pass by the various organs which are in its midst. What its mode of circulation really teaches, I will speak of hereafter, when we have seen its operations in other infusorians. The contractile vesicle (*cv*) is a simple, small, globular cavity, in the very middle of the body, and half-way between the top and bottom of the bell. It contracts three times in a minute, without exhibiting anything of the vivacity which is noticeable in many of the Infusoria, but simply lessens to an almost invisible point, and then as moderately expands to its full rounded contour. The reproductive organ (*n*) lies in a half circle across the body, on the ventral side, and at a little higher level than the contractile vesicle. Its tawny yellow color, and worm-shaped figure, render it by far the most conspicuous organ of the body.

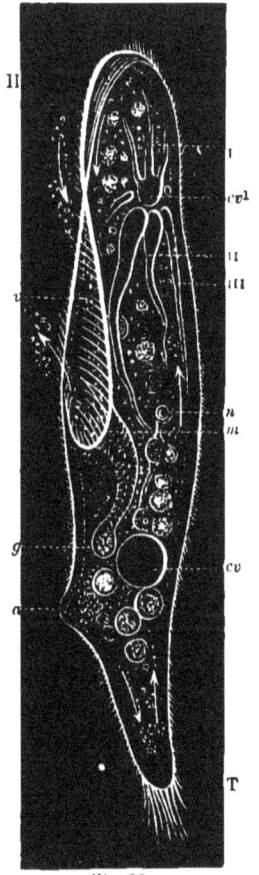

Fig. 96.

One of the closely allied kinds of Vorticellæ, a Zoöthamnium, exhibits the obliquity of the bell (fig. 104) much more clearly than the Epistylis, and especially so because I have drawn it as it appeared in profile; a position in which all of the bell-animalcules show the greatest degree of obliquity.

As we pass to the next illustration, we will take a glance at *Stentor*, (fig. 30,) merely to observe that the obliquity of its figure is very decided when compared with that of Epistylis or Zoöthamnium, and most emphatically so when it is swimming in a semi-expanded state (fig. 31). It will also serve as an intermediate step in the passage to the more oblique Infusorians, such, for instance, as we have represented in this figure, (fig. 96,) a Paramecium.

Fig. 96. *Paramecium caudatum.* Ehr. 340 diam. A view from the dorsal

*Paramecium* is one of the most common of its class wherever stagnant water or substances decaying in water exist. As it is so readily obtained, and so easy to observe on account of its size, and because it has afforded me the best opportunity of becoming acquainted with the structure of Infusoria, I shall take the liberty of describing its organization with considerable detail. Its figure may be compared to an elongated oval, with one end flattened out (H) broader than the other, and twisted about one third way round, so that the flattened part resembles a very long figure 8. The latter corresponds to the disc of the Stentor, Epistylis, &c.; and in accordance with this we find the mouth ($m$) opening near its edge, at a point which is halfway between the two ends of the body. I hardly need repeat to you that the disc is in the front region of the animal. Taking this now as a basis, it is not difficult to see the resemblance to the disc-region of Stentor, (fig. 30, $s$,) and that the moderate obliquity and twist of that of the latter are only extended to a much greater degree in Paramecium without disguising or changing its character in the least. The posterior (fig. 96, T) half of the body of Paramecium then corresponds to the slender part of the trumpet-shaped Stentor. A foreshortened view of the anterior end (fig. 97) of the former bears a still more striking resemblance to the disc of Stentor. In both we have the mouth (fig. 30 and fig. 97, $m$) at the bottom of a broad notch or incurvation, and the contractile vesicle ($cv^1$) on the opposite side, next the convex back, whilst the general cavity of the body lies between these two. Confining myself now to Para-

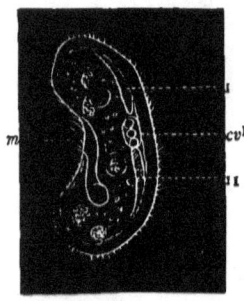

Fig. 97.

side. H, the head; T, the tail; $m$, the mouth; $m$ to $g$, the throat; $a$, the posterior opening of the digestive cavity; $cv^1$, the anterior, and $cv$, the posterior contractile vesicles; I, II, III, the radiating canals of $cv^1$; $n$, the reproductive organ; $v$, the large vibrating cilia at the edge of the vestibule. — *Original*.

Fig. 97. The same as fig. 96, seen endwise; *i. e.*, foreshortened from the head backwards. Letters as in fig. 96. — *Original*.

mecium, I shall proceed to illustrate its organization, and leave any further comparison with that of Stentor to your own inclinations. This, I think, will not be found very difficult to make.

The mouth (fig. 96, $m$) of Paramecium opens, as I have said, at the bottom of an excavation on the abdominal side, and from it the broad funnel-shaped throat ($m$ to $g$) passes into the body toward its left side in a curve which is continuous with the spiral trend of the disc, and terminates in a slightly rounded expansion ($g$) half-way between the mouth and the posterior end. At first sight, this is all there seems to be that is definite in character about the digestive system, and what I am about to relate now of my experiments in feeding the animal would appear to confirm this conclusion. Yet, before I get through, I hope to set the matter in a very different light. The intussusception of food by this infusorian is a remarkable phenomenon in its ordinary course of operations, but it is heightened exceedingly by supplying the creature with some strongly colored substance which is palatable to it. I have used indigo and carmine, both of them substances in a partially decomposed state; the indigo being the settlings from the maceration of the leaves of the Indigo plant, and the carmine the dried Cochineal insect. I found the indigo to be the more available on account of its dense color. A single drop from a common indigo-bag is sufficient to furnish occupation for hundreds of individuals. Where the indigo is collected in little heaps, under the microscope, the Parameciums gather around in swarms; and usually remain so quiet whilst feeding that any single individual may be selected for observation, and kept under the eye for a long time, — sometimes half an hour. I have represented by arrows the course of the particles of indigo as they are whirled along, by the large vibrating cilia ($v$) of the edge of the disc, against the vestibule of the mouth. Those which are accepted are passed down the throat, by the action of minute vibratile cilia, until they reach the rounded termination, ($g_{,}$) and there they are kept in a constant state of rapid revolution until a pellet of comparatively considerable size is formed, and enveloped in a sort of jelly, or some such transparent substance. When this is complete, the prepared morsel glides through an invisible outlet into the seemingly simple cavity of the body. If, now, the course of these pellets is carefully watched and noted, we must conclude that they are subjected to the action of something which has a definite position and a corresponding effect upon them, for they do not float about hither and thither indiscriminately, but follow each other in regular succession; passing from the throat backwards along the left side to the end of the body, they there turn forwards in a stream close to the back, and at the same time gradually move over to the right side, until they reach the head, (H,) and then, making a second turn, they follow the ventral line of the body past the mouth to a point which is a short distance behind the place where they left the throat, but considerably to one side of it, in fact, just about half-way between the right and left of the body. During this circuit, they gradually lose their jelly-like envelope, and the minute particles of indigo are, one by one, separated from the pellet and subjected to the process of digestion until whatever nourishment was in them is extracted, and then they cease their circulation and gather at the point which I have just mentioned. There they wait until a considerable mass of *rejectamenta* is formed, when the prominence, ($a$,) which has been formed a short time before, opens at the end and allows it to pass out, and then subsides and leaves not the least trace of an outlet.

Why, therefore, I would ask, ought we not to attribute to Paramecium a definite digestive or intestinal canal, seeing that the food passes in such a strictly confined channel? How well marked the boundary of this canal is, can only be inferred from the spiral course which the food pursues in passing from the mouth to its orifice of expulsion. This might be compared, by some, to the circulation of fluids in the cells of plants; but, from numerous and extended observations of my own, not only upon the familiar and well-known examples, such as Vallisneria, Naias, Udora, Chara, Nitella, Zygnema, the hairs of the Nettle, Tradescantia, and Circea, but also upon the cells of the pulp of various fruits, and upon many microscopic plants, I know that where there are distinct currents running in tortuous channels they constantly change their course, and by that evince their freedom to move anywhere and at any time throughout one common cavity; whereas in the infusorian before us the line of progress of the moving pellets, though a winding one, is fixed within definite limits, and clearly under the control of some guiding power. I have already pointed out the enormously elongated throat of Epistylis, (fig. 95, $g$ $g^1$,) but I revert to it for the sake of comparing it with the short one of Paramecium, and to show by the extent of the former that it is not impossible that its length in the latter may be carried equally as far, if not farther, although in a less recognizable form.

In this connection, I would like to say a few words in regard to the celebrated theory of Ehrenberg, because I think that those who have so positively denied the presence of the least trace of the little saccular stomachs which that eminent observer has asserted the Infusoria possess, are as far in the wrong as he, and perhaps more so. I have copied one of the illustrations of Ehrenberg, which he intended should exhibit the position of the stomachs, (fig. 98, $s$, $s^1$,) and their relation to the throat and mouth ($m$). He asserts that by feeding the infusorians with indigo or carmine the particles of colored matter may be seen to enter the mouth, pass along the throat, and, entering the cavity behind it, proceed to lodge themselves in the little globular saccules or pouches, ($s$, $s^1$,) which appear here to be attached to a central sac or canal. 

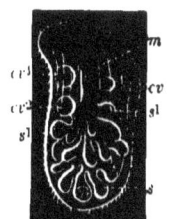

Fig. 98.

Now it must be apparent to you that such a configuration of the digestive system is altogether incompatible with the mode of circulation of the food as I have described it in Paramecium and Epistylis. Yet, under certain conditions, I have seen what appeared to be the same as that which the Berlin naturalist described; but it seems to be explained by a

Fig. 98. *Chilodon cucullulus*. Ehr. 200 diam. $m$, the mouth; $s$, $s^1$, the so-called stomachs; $cv$, $cv^1$, $cv^2$, the three contractile vesicles. — *From Ehrenberg*.

different interpretation of the relation of parts, and yet not detracting in the least from the reliability of Ehrenberg as a microscopic observer.

Suppose, for instance, that a solid body were hollowed out in such a way as to leave numerous projections pointing toward the centre, the effect when seen in profile would be to give it the appearance of being divided part way across by numerous partitions which separate as many little side-chambers, and thus you would have the same aspect presented as that represented here, with this exception, that in the *Chilodon* (fig. 98) it is, according to Ehrenberg, a sort of bag whose thickness is puckered or drawn out into little side-pockets, and the whole suspended within the body of the animalcule. As I said, the appearance is nearly the same in both; but the possibility for the circulation of the food is far more easily conceived in the supposititious body than in the other; and such do I believe to be the true state of the digestive system, rather than as our author has avowed. By Ehrenberg's theory, the pellets of food would have to pass from one pouch toward the centre of the body and then outwards into the next pouch, and then again in the same way into the next and the next, and so on in a very tortuous course, all along one side of the body backwards and along the other side forwards, in order to complete such a circuit as you have had illustrated in Paramecium; whereas in this latter it progresses in an even tenor, almost insensibly gliding, from the right to the left, from one end of the body to the other,— and yet this could be done notwithstanding the presence of the above-mentioned projections which I believe to lie across the path, and which falsely seem to be the walls of numerous side-chambers.*

The system which is analogous to the blood-circulation of the higher animals is represented in Paramecium by two contractile vesicles, (fig. 96, $cv$, $cv^1$, I, II, III,) both of which have a degree

* By killing this infusorian with a drop of the extract of opium, the effect is to partially condense the circulating contents, when the same appearance is presented as Ehrenberg has represented for the form of the digestive system of *Trachelius ovum*; but the lateral branches are much more numerous in the former.

of complication which, perhaps, exceeds that of any other similar organ. I have represented them in this figure in two of their most striking conditions; such as you would hardly suppose could be exchanged the one for the other; and yet for hours have I watched them, — as clearly visible as they appear to your eyes in the illustration, — alternately assuming these two phases. They both lie close to the thick wall of the body on the side nearly opposite to that in which the mouth opens, but their canals extend almost around to the latter point. When fully expanded ($cv$) the main part of each vesicle is about half as broad as the body, and there is little or no trace of the canals to be seen; but the moment contraction begins, they appear as fine radiating streaks, and as the main portion lessens they gradually broaden and swell until the former is emptied and nearly invisible, and they are extended over half the length of the body. In this condition they might be compared to the arterial vessels of the more elevated classes of animals, but they would at the same time represent the veins, since they serve at the next moment to return the fluid to the main reservoir again, which is effected in this very remarkable way. As the main vesicle ($cv^1$) begins to expand, the canals, one after the other, and in nearly regular sequence, suddenly open into it at their broad end ($\text{II}$) with a sort of puff, or as if a valve had been jerked back, and, a communication being thus established again ($\text{I}$) with the central point, their contents are quickly poured out, and then in like succession they contract until nothing is to be seen of them, whilst the vesicle expands to its fullest extent, — as we see it ($cv$) in the posterior half of this figure. The contents of these vesicles is a perfectly clear fluid, without the least trace of granules, or anything that might correspond to the blood corpuscles of those animals which possess a true circulatory system.

The reproductive organ is excessively faint, and very difficult to detect, even with the best powers of the microscope, except at certain seasons when the eggs are full grown. It consists of a slender tube in which the eggs ($n$) are placed in a single line, one after the other, at varying distances. It usually lies in the midst of the body, about as far from one side as the other, and extends from one half to two thirds of the length of the animal. According to Balbiani's observations upon a closely allied species, when the eggs are laid they pass out from the ovary through an aperture near the mouth.

The organs of locomotion of Paramecium are mostly very short, closely-set, vibrating cilia, which completely cover the body, and are arranged in numerous longitudinal lines. At the tail (T) there are quite a number which are much larger and longer than the others; and in the vestibule (v) of the mouth (m) they have more of the character of lashes, on account of their large size and frequent flexions toward the latter as they assist to throw off an unwelcome morsel of food, or enforce the entrance of some vagrant bit which may happen to be too large for the efforts of the smaller cilia in the throat. The apparent intelligence with which one, two, or three of these lashes are used like so many fingers to turn over, or lift up and rearrange some difficult, more than mouthful of food, is very striking, when we consider the low grade in which the Infusorians stand among living beings. How often have I observed one of these creatures trying to get a great globule of food into the throat, and seen it whirl it first one way then another, and finally, after several efforts, the attempt being given up with a sudden, petulant rejection of the desired morsel, throw it away with a toss far from the influence of the ciliary vortex!

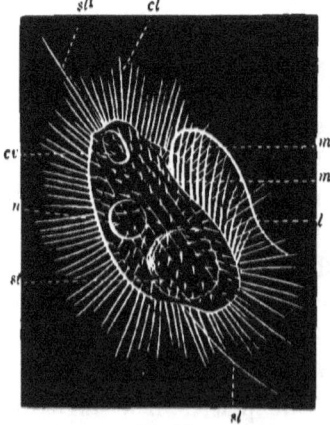

Fig. 99.

The next Infusorian which I shall make mention of is *Pleuronema;* one which I have already described (p. 148) pretty fully, and will therefore introduce here (fig. 99) merely to draw your attention to the fact that its vestibular disc (fig. 99, *m*) has a greater obliquity than that of Paramecium;

appearing from this point of observation more like a slit or furrow trending parallel with the side of the body. An end view of the animal has a strong resemblance to the disc of the trumpet-shaped Stentor (fig. 30); much more even than a similar view of Paramecium (fig. 97).

The extreme of obliquity in this type is best exemplified in a minute marine Infusorian which is called Dysteria (fig. 100). It has the appearance of anything but an Infusorian; and in fact such a strong resemblance does it offer, not only in form, but also in habit, to some of the articulated animals which are known as Rotifera, as to induce certain naturalists to believe that it forms a transitional link between the Infusoria and Articulata. A careful investigation of its structure, and the relation of its organs to each other, will convince the unbiased observer that it is in every respect a true Infusorian, and that its resemblance to the Rotifers is such as often occurs between animals otherwise totally different. It is an infusorian between two leaves or flexible shells (*h, t, up, lw,*) of unequal width, which are united by a sort of hinge along the left border (*h* to *t*), and gaping to a more than equal extent along the right side, where the upper one (*up*) far overhangs the other (*lw, bk*) throughout the whole length

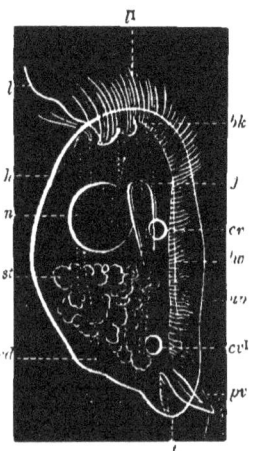

Fig. 100.

Fig. 99. *Pleuronema instabilis*, n. sp. 1000 diam. *m*, the mouth; *st*, the food gathered in one mass; *mc*, large cilia in the vestibule of the mouth; *l*, the single vibrating lash projecting from *m*; *cl*, vibratory cilia covering the body in rows; *sl*, the posterior, and *sl*$^1$ the anterior saltatory cilium; *cv*, the contractile vesicle; *n*, the reproductive organ. — *Original.*

Fig. 100. *Dysteria prorœfrons*, n. sp. 600 diam. A view from the broader side. *h, t, up,* the broader valve; *lw,* the right edge of the narrower valve; *bk,* the beaks of the narrower valve; *rd,* the ridges on the surface of the broader valve; *pv,* the pivot; *l,* the large, single vibratory lash; *l*$^1$, vibratory cilia; *j,* the jaws; *st.* the mass of food; *cv, cv*$^1$, the two contractile vesicles; *n,* the reproductive organ. — *Original.*

of its free edge. The broader or dorsal shell (*up*) is convex toward the eye, and the whole organization lies within its concavity, whilst the narrower one (*bk*, *lw*) is flat, simply covering the body, and as a natural consequence does not include any part of it. The open space between them is endowed with a row of closely-set, large, vibratile cilia, which differ in size according to their position; those in front (*l¹*) being by far the longest, and those along the side scarcely more than half as long; and in addition there is one (*l*) which, from its great size, has more of the character of a proboscis, and is attached nearly at the extreme anterior border of the row.

It is not an easy matter in this case to determine how much of the one-sided, cilia-bordered furrow corresponds to the disc or vestibule of Epistylis, Stentor, Paramecium, or Pleuronema; nor does it affect the question of the degree of obliquity of the conformation of this animal, as long as we see that, whatever it may be, either wholly or in part a vestibule, it is at least extremely oblique, and that it is not possible to view it from any point but that the body appears asymmetrical in relation to it.

The most striking peculiarity of this creature is its habit of swinging around on a pivot, (*pv*,) which consists of an ovate or lancet-shaped appendage, of considerable dimensions, that projects from near the posterior end of the body, and in the line of the row of cilia. The pivot possesses perfect flexibility at its base, so that the animal can move over a considerable distance backwards and forwards without disturbing the point. Most of the time it keeps the flat side down when gyrating around its place of attachment, but now and then it turns up on its right edge, and performs its eccentric rotations about the appendage. This is the habit which, as I said before, has impressed some observers with its similarity to the Rotifera. In connection with this, too, it happens that the creature possesses a pair of jaw-like, or rather pincer-like bodies (*j*) which lie near the entrance to the mouth, and occasionally open and shut like a pair of forceps, just as similar bodies known as the jaws of Rotifers do, whilst food is passing between them. Excepting the passage between these jaws, there is not the least trace of an intestine, nor of any definite cavity devoted to digestion. The food occupies the whole length and breadth of the body under the same circumstances as are observable in Paramecium, Pleuronema, Stentor, &c.

The contractile vesicles are two ($cv$, $cv^1$) quite small globular bodies, one of which is situated just to the right of the jaws, and the other close to the base of the pivot; and although they contract very slowly, not oftener than once in four or five minutes, they evince every characteristic, in action and physiognomy, of true infusorian, pulsating vesicles. The large colorless reproductive organ ($n$) singularly exemplifies in itself the one-sidedness of the animal by its conformation to the shape of the body. One side of it is convex and, like the rest of the organization, projects into the concavity of the larger shell, whilst the other face is flat and, as it were, moulded upon the plane shell. It forms a very conspicuous object just to the left of the jaws, and might easily be mistaken at first glance for a contractile vesicle, especially as the true representatives of that organ are so very inconspicuous both in regard to their size and actions.

Now in all the organization of this animal there is nothing which is not strictly infusorian in character. The jaw-like bodies ($j$) are not confined to this alone, for there are quite a number of others which possess a similar apparatus at or near the mouth. The Chilodon, which I have copied to illustrate Ehrenberg's idea of the nature of the digestive system, has a complete circle of straight rods (fig. 98, $m$) around the mouth. As for the pivot, (fig. 100, $pv$,) it is nothing but a kind of stem, such as exists on a larger scale in Stentor, or is more peculiarly specialized in the pedestals of Epistylis (fig. 95, $p$), Zoöthamnium (fig. 104, $p$), or Podophrya (fig. 25); and as counter to what we see in these last, I would state that there are certain of the Vorticellians, closely related to Epistylis, which have no stem whatever, and swim about as freely as Dysteria.

What characterize them all, are not only the oblique relations of right and left, but also the presence of one or more peculiar contractile bodies, the so-called contractile vesicles, and a diffuse digestive system. Perhaps now you may call to mind some of the Protozoa — for instance this animal, (fig. 25, p. 51,) a Podophrya, which, as it appears at this stage of growth, is an apparently symmetrical, four-sided, inverted pyramid — in which you cannot discover any signs of that obliquity of which so much has been said; and you will be induced therefore to set them down as exceptions to the rule. Now this apparent discrepancy has been one of the stumbling-blocks of naturalists, and has thrown some of them at least off the true course of development; and like the stragglers of a grand army, although they follow mainly in its line of march, they scatter about here and there, picking up anything and everything they can get hold of.[*] Now as I do not wish to anticipate anything, I will defer the removal of this stumbling-block until another occasion.

My present aim is simply to explain the relation of the *type*

[*] See the sensational assertions of L. Agassiz in regard to the systematic position of various Protozoa; for instance, in the "Proceedings of the Boston Society of Natural History," vol. III. Nov. 6, 1850, p. 354, he declares that the well-known parasites (*Trichodina*) of Hydra are Medusæ. "In the eggs of Hydra, he had been able to trace all the forms from a segmented yolk to these parasites; the fresh-water Hydra is the Polypoid form of Medusæ, while these parasites are the Medusoid form." Of the same degree of reliability is his assertion in his "Essay on Classification," Boston, 1857, p. 182, or the London edition, 1859, p. 291, where he says, "I have seen, for instance, a Planaria lay eggs out of which Paramecium was born, which underwent all the changes these animals are known to undergo up to the time of their contraction into a chrysalis state; while the Opalina is hatched from Distoma eggs." Or "Essay," &c., p. 75; Lond. ed. p. 112: "Having satisfied myself that Colpoda and Paramecium are the brood of Planariæ." Finally, to complete the disjointure of this homogeneous group, and its entanglement with Articulata through Opalina, and with Medusæ through Trichodina, he asserts, in another part of that same essay, that the Vorticellidæ, which were well known, long before the time he wrote, to be the crowning rank of Protozoa, "differ entirely from all others," (p. 182; Lond. ed. p. 290,) *i. e.*, such as Stentor, Paramecium, and the rest of the Enterodela, and that he had satisfied himself of the "propriety of uniting the Vorticellidæ with Bryozoa," (p. 72; Lond. ed. p. 108). While I am about it, I would venture to guess that that author's pretended Hydroid animals of Millepora are mere *parasites!*

of this division to *bilaterality*. I will endeavor to show how the obliquity of the spiral is resolvable into a bilateral form. A strip of paper shaped into a long parallelogram and folded

Fig. 101.

Fig. 102.

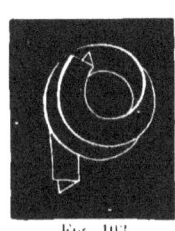
Fig. 103.

lengthwise along the middle will represent the idea of bilaterality, one half forming the right and the other the left side. Now,

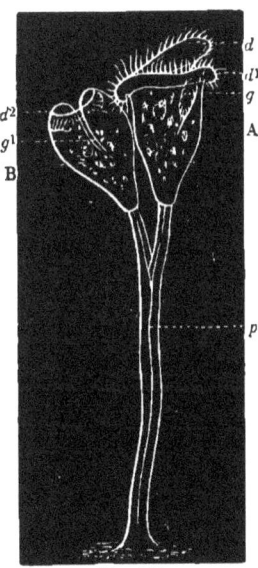
Fig. 104.

for the sake of convenience in winding it up, cut it lengthwise along the middle, (fig. 101,) except for a short distance at one end, which may represent the *head*. Then, the two halves being folded together as before, it may be wound up in a spiral with the uncut end, i. e. head, on the outside. This may be wound either to the right or to the left, according to circumstances. When coiled in a long-drawn-out spiral, (fig. 102,) we have the form of Dysteria typified, or, twisted a little closer, a Spirostomum, then still more, a Paramecium and Pleuronema, when made into a nearly flat spiral, (fig. 103,) a Stentor, and finally, when wound up

Fig. 101. *Zoöthamnium arbuscula*. Ehr. 200 diam. A stem with two individuals. A is fully expanded; B is partially shut up, by the infolding of

# THE IDEAL PROTOZOAN AND BILATERALITY.

like a watch-spring, Epistylis and Zoöthamnium. Yet in all these various degrees of the twist of the spiral, you do not lose sight of the two halves, i. e., its *bilaterality*. Nor is this altogether an ideal transition of degrees of obliquity; for I have seen one and the same individual assume widely diverse states of this kind, — as illustrated in these two figures, (figs. 104 and 105,) — and pass from one into the other whilst I was observing it. When I first saw it, the form was like this, (fig. 105,) in which the edge of the disc is drawn out and twisted, (*d*,) as you would a ribbon, into three or four coils; but gradually this proceeded to unwind, until, in about ten minutes, it had straightened out and projected above the rest of the disc like a broad lip, (fig. 104, *d*).

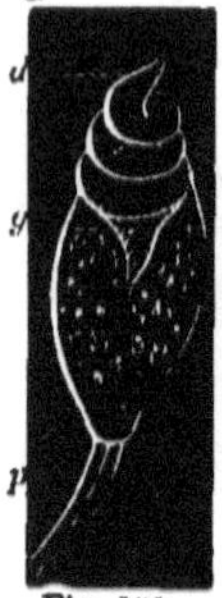

Fig. 105.

the edge of the bell; *p*, the stem, with a thin thread-like muscle in the axis; *d*, the lid-like portion of the disc; *d*$^1$, the edge of the disc; *d*$^2$, the edge of the disc rolled in; *g*, the vestibule of the mouth; *g*$^1$, the throat. — *Original*.

Fig. 105. The same as fig. 104, with the edge of the lid-like part of the disc drawn out and coiled into a spiral; *d*, the point of the spiral; *g*, the vestibule of the mouth; *p*, the stem. — *Original*.

# CHAPTER X.

### ZOÖPHYTA.

I PROPOSE to take up next the group of *Zoöphyta*, of which the Sea-anemones, (fig. 28, p. 57,) Corals, Hydras, (fig. 27, p. 55,) Jelly-fishes, (fig. 37, p. 70,) Starfishes, (figs. 109, 110,) Trepangs, (fig. 54, p. 120,) and the like, are there presentatives. This is a type of organization in which the various organs repeat themselves, more or less, between the back and the abdominal mid-line of the body; that is to say, they are *laterally repetetive* on each side of an imaginary plane which divides the body exactly into right and left halves. Whether the repetitions are few or many, they always have reference to the two halves of the body. This multiplication of parts I have already pointed out in the organization of the Sea-anemone, (pp. 57, 58, figs. 28, 29,) showing that its semi-partitions, as it were, split the body lengthwise into so many galleries. I will not repeat what I have already said in regard to this animal, but will add a few words in reference to the mode of proceeding which is adopted when the parts of the organization are multiplied as the individual increases in age. Exteriorly this is most conspicuous in the development of the feelers or tentacles. As I have already said, (p. 59,) the young begins with six double semi-partitions, which correspond in position to the first six tentacles; but what I will point out here is, that the tentacles always agree with the number of the *pairs* of the partitions, not only in the young but also in the largest individuals, where these feelers may be counted by hundreds.

At first the young Anemone develops the primary set of tentacles, (figs. 106, 107, I,) and then the semi-partitions $(p, p^1)$ appear next, and in such a position as to seem to be inward prolongations of the wall of the hollow tentacles; so that the space between each pair is in direct continuation with the

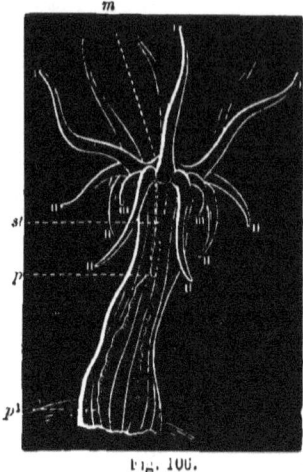

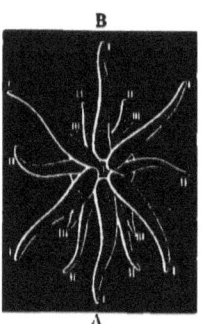

Fig. 107.

Fig. 106.

cavity of each feeler. The second set of tentacles (figs. 106, 107, II) originate at alternate points with the first six, (I,) but behind them; and oftentimes, as if to exemplify their distinctness from the former, they are held in a different position, turned back, for instance, as represented in this figure, (fig. 106, II). The third set (III) alternate again with both the first and second, and therefore are twelve in number; and originate still further back from the mouth. In these figures they are as yet quite small, (III,) and you can see but a part of the set here, because, as I have said on a previous occasion, (p. 60, note †,) they do not all develop at an equal rate. When, now, the number of tentacles has increased by another set, which adds twelve more, there are then twenty-four in all, and we find no difficulty in ascertaining their relations; but upon the next

Fig. 106. *Metridium marginatum.* M. Edw. 6 diam. A profile view of a very young individual, seen in the direction A, B, of fig. 107. *m*, mouth; *st*, stomach; *p*, bottom of *st*, where it opens into the general cavity; *p* to $p^1$, the free edge of the semi-partitions; I, II, III, the different sets of tentacles. — *Original.*

Fig. 107. A foreshortened view of the head of fig. 106. A, B, the median plane which divides the body into right and left halves. — *Original.*

doubling they are so crowded as to render even the counting of them not a little troublesome, to say nothing of defining the exact relations of each one to its fellows. But I need not go beyond this point, however, as sufficient has been said to show how this method of increase is carried out.

I will merely reassert what I have already said, now that you are more familiar with the organization, that these successive alternations of the series, one, two, three, four, &c., of tentacles and their corresponding partitions, always have a direct reference to the right and left sides of the body. This you may recognize in the diagram (fig. 108) of the successive productions of the partitions, much more clearly than could be seen in the crowded disc of an equal number of tentacles. If you attempt to divide the body into equal halves by running a line through any other two opposite points than those which lie in the prolonged vertical plane of the flat stomach, (I, II,) you will

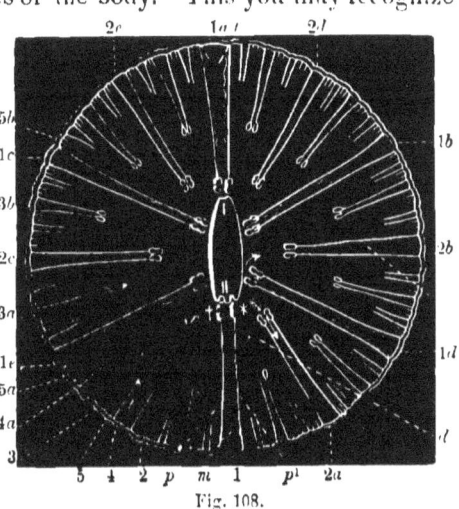

Fig. 108.

Fig. 108. *Cereus Sol.* Verrill. 2 diam. From Charleston, S. C. A foreshortened view of the interior, from just behind the mouth, backwards to a point a little behind the posterior end of the stomach (I, II). *d*, the general digestive cavity; *l*, spaces between the pairs of partitions, the arrows indicating the passage-ways to and from the general cavity (*d*); *p*, *p*¹, the two partitions of one of the largest pairs; *m*, the muscular layer on the opposite faces of *p*, *p*¹; *.* †, the reproductive organs, which in profile (fig. 28) appear like a deep frill along the edges of the partitions, extending from the posterior end (fig. 28, P) of the body toward the head; I, II, the two plicated, upper and lower, edges of the flat stomach; 1, 1*a*, 1*b*, 1*c*, 1*d*, 1*e*, the six pairs of the first set of partitions; 2, 2*a*, 2*b*, 2*c*. 2*d*, 2*e*, the six pairs of the second set; 3, 3*a*, 3*b*, &c., the twelve pairs of the third set; 4, 4*a*, &c., the twenty-four pairs of the fourth set; 5, 5*a*, 5*b*, &c., the forty-eight pairs of the fifth set. — *Original.*

utterly fail in your undertaking; and the conclusion will inevitably be that *bilaterality* is the dominant feature here, to which all other arrangements are subservient and *secondary*. But this is not all that we have yet to learn from the study of the *taxis* of this animal. Not only does the plane of bilaterality impress its character upon the parts of the organization, so that they are, as it were, dependents upon its two faces, but the two edges of this plane have diverse significations, as clearly demonstrable as that the plane which divides man's body into right and left has one edge which corresponds to the back and the other to the front. You may see this character even in the youngest Anemone, where of the two tentacles opposite the two ends of the oblong mouth one (fig. 107, A) is larger than the other; but it is most especially exemplified, in the more advanced stages of growth, by a peculiar fold at one corner of the mouth which is continued within along one edge (fig. 108, II) of the flat sac which answers for a stomach.

Certain foreign species of Anemones have this fold very highly developed; and in one in particular it projects like a great trumpet, as far out as the tentacles reach.[*] Which of the two it is, the back or the front, that this fold corresponds to, naturalists are not agreed upon; but as the discoverers of the Anemone with the trumpet-like fold state that its food is passed into the mouth down this funnel, we may infer from analogy that the side of the body in which it lies is the predominant one, and therefore is entitled to the designation of the *front*, or ventral side. But still the determination is not an absolute one, nor is it necessary for our purpose that it should be so; suffice it to say that we can recognize two antagonistic sides above and below a horizontal plane which divides the body into an upper, or dorsal, and a lower, or ventral region.

Thus in those Zoöphytes which, from the comparative simplicity of their organization, and the manifold lateral repetition of parts, must be looked upon as the lowest of the grand divis-

---

[*] *Siphonactinia Bœckii.* Danielssen and Koren. Fauna Littoralis Norvegiæ, 2$^{de}$ liv. 1856, p. 88, Pl. XII. figs. 4, 5, 6.

ion to which they belong, the dominant character of the animal kingdom, *bilaterality*, is as clearly laid down as in any of the higher types; and at the same time the upper and lower sides are indicated to that extent and to that degree of specialization which will not allow of a symmetrical bipartition of the body in any other way than that which I have explained to you. I would also mention, in passing, that, in the most remote geological ages, the earliest corals that appeared on earth were as distinctly characterized by a right and a left and a front and a back as are those of the present day.

As I shall have occasion hereafter to speak of the next higher class, to which the jelly-fish, (fig. 37,) sea-blubber, hydra, (fig. 27,) &c., belong, I will pass over it now, and also another closely related class, and lay before you the organic relations of the Echinoderms, — the fourth and highest group of this division.

*Echinodermata.* I have selected for illustration two of the

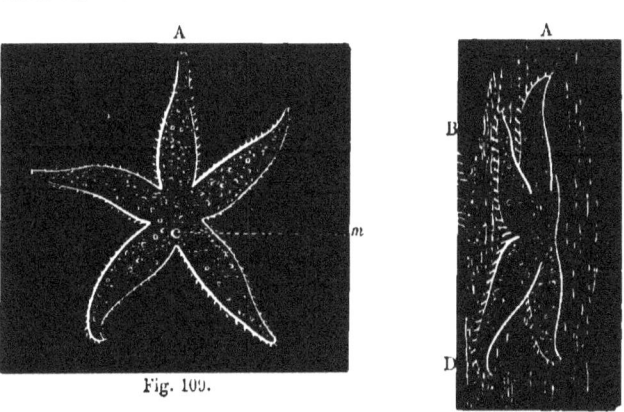

Fig. 109.

Fig. 110.

most widely diverse exemplifications of the lateral repetitive type

Fig. 109. *Asteracanthion rubens.* M. and Tr. The common Starfish. One fourth natural size. A view of the posterior face. $m$, the madreporiform body. A line drawn through A, to the madreporiform body ($m$), divides the body into right and left halves. — *Original.*

Fig. 110. A semi-profile view of fig. 109, as seen when creeping up the perpendicular face of a rock by means of its tube-like feet, or suckers. A, the median arm; B, D, the arms of one side. — *Original.*

among the members of this class: the one a Starfish, (figs. 109, 110, and 111,) in which the multiplication of parts is carried out to the greatest degree, and the other, one of the Trepangs, (figs. 114, 115, 116,) in which the repetitions are the least in number. In these two foreshortened views (fig. 111 and fig. 115) of the animals in question, you may see at a glance the difference in the degree of repetition; but you will understand better what organs are affected thus in both, after I have illustrated them from the most natural point of view. I merely draw your attention to these views now, because one of them, the Starfish, usually presents itself to the observer in this position, (fig. 109,) on account of its peculiar form, which can hardly be said, in common parlance, to have a profile view. When we speak of

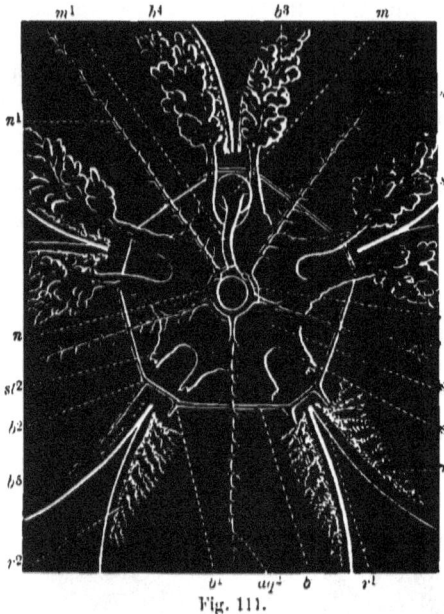
Fig. 111.

Fig. 111. *Asteracanthion rubens*. M. and Tr. A diagramic, foreshortened view of the anterior face, to exhibit the relations of the various organs. The median arm is *below*, and the madreporiform body ($m$) is in the middle line above. The ends of the arms are left out of view. $st$, the stomach; $st^1$, the puckered prolongations from $st$; $st^2$, $st^3$, the bases of $st^1$, the latter being cut away; $aq$, the aquiferous ring about the mouth; $aq^1$, $aq^2$, the aquiferous canals; $m^1$, the madreporic canal; $m$, the madreporiform body; $b$, the so-called circular canal of the blood circulation; $b^1$, $b^2$, the points where $b$ gives off canals ($b^5$) to the arms; $b^3$, $b^4$, branches from such as $b^5$, which spread over the puckered prolongations from the stomach; $n$, the nervous ring about the mouth; $n^1$, $n^2$, branches from $n$; $r$, $r^2$, the reproductive organs; $r^1$, the point where the eggs escape. — *Original.*

the length of a tree, we mean along a line drawn from one end towards the other; but if now one were to cut across the trunk and take out a transverse slice as thin as a wafer, it would seem almost-absurd to speak of the length of the slice, meaning its thickness; but yet the term would appear perfectly justifiable to any unprejudiced mind upon considering its relations to the surrounding parts from which it was taken: and so it would be in relation to the starfish. Its length is commonly called its thickness; but I could show you other starfishes which are nearly or fully as long, or, commonly speaking, as thick, as they are broad; and among the fossils certain starfishes (figs. 112, 113) which, when alive, were attached to stems, and whose bodies were much longer than broad; and yet, in a view from the mouth end, (fig. 112,) you might think you were looking at one of the living starfishes, foreshortened by the plastic hand of Nature.

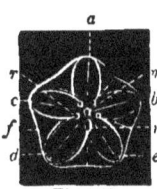
Fig. 112.

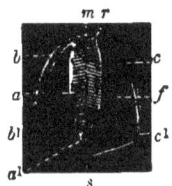
Fig. 113.

I now refer you to this side view (fig. 114) of the Trepang, and then to its foreshortened aspect (fig. 115), which may be compared to the thin, transverse slice from the tree, and ask you if it were possible to discover its length in any other way than along the line of vision, that is, right through what I might call the thickness of this figure?

We will, then, survey the interior of the starfish as if we were looking into the mouth of a short tube and saw five groups

Fig. 112. *Pentatremites florealis.* Say. Natural size. A foreshortened view of the anterior face. $a$, the median arm; $b$, $e$, the two left arms; $c$, $d$, the two right arms; $f$, the region between the arms formed by their lateral conjunction; $m$, the mouth; $r$, the right and left openings of the reproductive organs; $r^1$, the cloacal cavity, forming the common point of opening for the posterior end of the intestine and a part of the reproductive organs. — *Original.*

Fig. 113. The same as fig. 112, in profile, as seen from above. $a$, the median arm; $a^1$, end of $a$; $c$, $c^1$, the right, and $b$, $b^1$, the left arms; $f$, the same as in fig. 112; $s$, the posterior end of the body, where the stem is attached; $m$ and $r$ as in fig. 112. — *Original.*

of similar bodies so disposed within it that one group trends along its lower central line, and the others are arranged right and left, two on one side and two on the other. Let us see now of what these groups consist. In the centre is the mouth, (fig. 111,) which is a simple round aperture, that leads, through a short throat, into a broad stomach ($st$), which may be described as having five sides, namely, one on the abdominal, or lower line of the tube, which corresponds to one of the arms of the starfish, and two on each side at equal distances apart, opposite the other four arms. From the five corners of this organ five pairs of deeply folded and puckered sacs ($st^1$) stretch out into the arms, nearly to their tips. Each pair is divided between two adjacent arms, so that every two of each arm belong to different pairs. From the mouth into the stomach, and thence to the ends of the ten plicated sacs ($st^1$) is one unbroken communication, the whole expanse of which is devoted to digestion of food; the saccules performing the part, it is thought, which corresponds to what is called in the higher animals chylification; a process which usually succeeds the first rough preparation by which the particles of food are reduced to a semifluid state.

At a short distance behind the mouth the throat is encircled by two rings, ($aq$, $n$,) of which the larger one ($n$) is nearest the eye, as we now survey the organization. From the latter, which is looked upon as a *nervous collar*, five equidistant nervous threads ($n^1$, $n^2$) extend along the middle line of each arm, and, as they pass, give off, in opposite pairs, numerous minute sidebranches to the neighboring organs, and especially to the double rows of tube-like feet (see fig. 110) which project through a series of pores along the front face of each arm.

In like manner from the smaller ring, ($aq$,) which is hollow, five equidistant, thin tubes ($aq^1$, $aq^2$) arise, and run along the middle of each arm; but as each immediately underlies a nervous thread, the lower median nerve and the next one on its right are left out of the illustration in order to expose some of them in this figure. These tubules also give off opposite pairs of branches, which penetrate the tube-like feet that I just mentioned.

OF STARFISHES. 185

The main feature, though, of this *fluid circulating system*, is the large canal ($m^1$) which extends from the upper side of the circular tube ($aq$) and backwards along the superior median line of the body, and terminates in a trumpet-like dilatation ($m$). On the outside of the body the mouth of the trumpet is covered by a circular, stony button which is peculiarly ornamented by labyrinthine furrows, that give it the appearance of a minute brain-coral. On this account it has been called the *madreporiform body*. It is readily detected, in our common starfishes, as a little, faintly rose-colored, hard, circular plate, (fig. 109, $m$,) from one eighth to one fifth of an inch in diameter, lying at the angle between two arms.

Now, in all the numerous and various kinds of Starfishes and Trepangs the trumpet-shaped body, ($m^1$,) which is called the *madreporic canal*, may be found in some guise or other at the median line of the body; and serves as a guide by which we may at a glance determine the right and left sides. How closely related in their arrangement the other organs are in reference to this one you will see presently in a most convincing illustration; it is, therefore, now spoken of in this light, that it may be referred to hereafter as the eminently prominent, material symbol of the bilateral idea, in the midst of such a great frequency of repetitions among the various organs which are more or less intimately connected with it. Through whatever other line you may divide the body than the one in which this is situated, you get an asymmetrical figure.*

---

* Lest some misapprehension or objection should arise in the mind of any one who may bring to his recollection the asymmetrical lungs of serpents, or the one-sided reproductive organs of birds, &c., &c., I would state that these are totally different cases. These organs were perfectly symmetrical in the younger stages of growth, and clearly originated upon a basis of bilaterality; so that any subsequent changes in their proportions could not affect the typical relations of the other organs, any more than would the opening of one eye and the shutting of the other. In the madreporic canal we have a sort of base-line upon which all the other organs are built up, and from which they are, as it were, an ideal outgrowth and dependency, — subservient and *secondary* to the influence which it exerts as the ideal of the "*primitive stripe.*" The well-known instances of ec-

In addition to this there is another system ($b$, $b^1$, $b^2$, $b^3$, $b^4$, $b^5$) of tubes which embraces the outskirts of the stomach, and sends off a minute branch ($b^5$) to each of its ten puckered, chylific sacs ($st^1$). It has been asserted by some observers that this system is connected with the other one by a vessel which runs along the madreporic canal ($m^1$) and forward to the canal ($aq$) about the throat; but of this there is a divided opinion among anatomists, some insisting that no such relation exists. This is all owing to the minuteness and delicacy of these tubes, and the consequent great difficulty of investigating their relations. It has, however, been rendered more probable that they are distinct systems, the one emanating from the region of the mouth being devoted to the circulation of a watery fluid, and the other forming, more probably, a blood circulation.

Last in the list of these manifold repetitions, the reproductive organs come under our notice. These are ten in number, in five pairs ($r$, $r^1$, $r^2$). They stretch along the sides of the arms, two in each, for a short distance, and, branching on one side, send off their minute twigs into the general cavity. They arise near the angles of the arms; and it is at these points ($r^1$) that each discharges its eggs in the breeding season. From here they gradually taper to the opposite extremity, ($r^2$,) and terminate merely as irregularly shaped tubules, of an apparently very simple organization.

We have, then, no less than five, or perhaps six, different sets of organs arranged along the sides of this foreshortened, quintuplicate tube, among which some of them exhibit a fivefold, and others a tenfold repetition. How, and by what consecutive gradations these parts are reduced in numbers as we rise in the scale of successively higher forms, until we meet with the most eminent, singularly worm-like Trepangs, I have not sufficient space here to enlarge upon; but I must not fail to point to one remarkable fact which keeps itself most prominent during, and as an accompaniment to, these changes, namely, that the reduction

centricity of the madreporic plate in certain Echini would be cited as an objection here by no one but such as would mistake bisymmetry for bilaterality.

progresses at a perfectly symmetrical rate, and always in reference to right and left. To such a degree, moreover, is this carried in certain Trepangs, that the repetitions are reduced to nothing so far as concerns the reproductive and blood circulating organs, there being but a single one of each; and the parts of such as the digestive organ and the water circulating systems are simplified in a manner that leaves them under such peculiar relations to each other, that, until within a very few years, most naturalists have been in the habit of looking upon them as possessing more of the characters of those of a certain group of worms; — or rather, they united these worms with the vermiform Trepangs. How far they were justified in doing this I shall bring up for discussion hereafter. At present I shall produce for illustration one of our native Trepangs, in which, so far as the group to which it belongs is concerned, the process in question attains the climax of perfection.

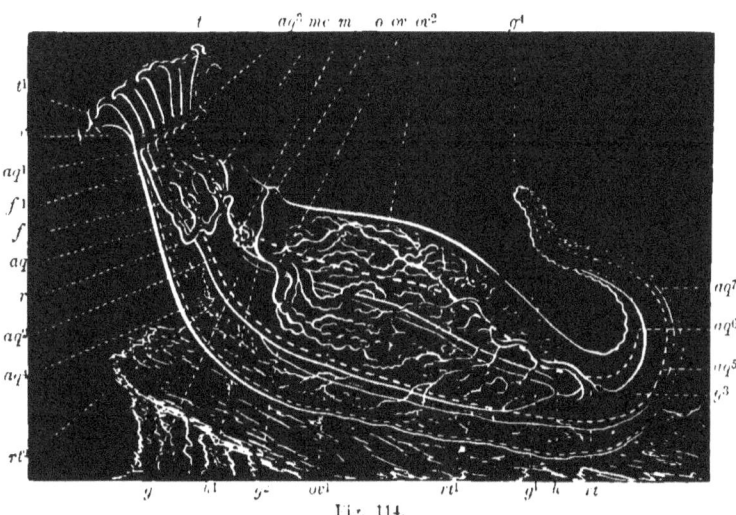

Fig. 114.

Fig. 114. *Caudina arenata.* Stmp. Natural size. A longitudinal, semi-diagramic view of a common Trepang of our coast. $t$, $t^1$, the four-pronged, anchor-shaped feelers of the head; $f$, $f^1$, the stave-like, calcareous, forked pieces of the buccal ring; $g$, the anterior end of the intestine; $g^1$, the first bend of the same; $g^2$, the second bend of the same; $g^3$, the posterior or cloacal region of the intestine; $g^4$, posterior aperture of the same; $rt$, $rt^1$, $rt^2$, the respiratory

Unlike that of the Starfish, the organization of *Caudina*, as this is called, is best understood when seen in profile; and I will therefore describe it from this point of view, (fig. 114,) and then compare its foreshortened aspect (fig. 115) with that of the Star-

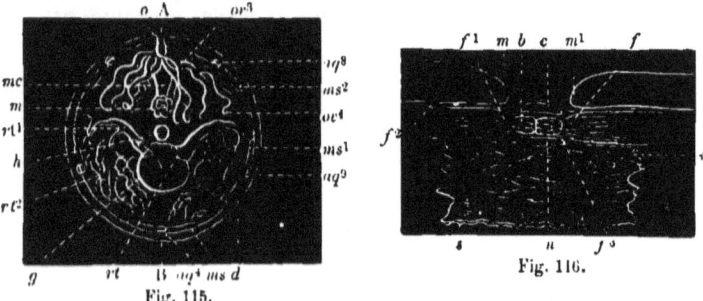

Fig. 115.    Fig. 116.

fish. The outline of this animal, as you see by the illustration before you, is that of a thick, worm-shaped figure, of which one end, the head, is rather abruptly terminated, as if cut straight

branches; $m$, the madreporic body; $mc$, the madreporic canal; $r$, the aquiferous ring; $aq$, the aquiferous canals going from $r$ to the space ($aq^7$) at the base of the feelers ($l$); $aq^1$, $aq^2$, $aq^4$, $aq^5$, $aq^6$, $aq^7$, the longitudinal aquiferous canals running close to the under surface of the skin; $h$, $h^1$, the heart; $c$, the ribbon-like nervous collar; $ov$, $ov^1$, $ov^2$, the reproductive organ; $o$, the external aperture of $ov$. — *Original*.

Fig. 115. A diagramic, foreshortened view of fig. 114. A line drawn through A, B, will represent the median plane of the body, dividing it into right and left; $g$, the intestine; $rt^1$, $rt^2$, the respiratory branches; $rt$, entrance to the respiratory branches; $aq^1$, $aq^8$, $aq^9$, the positions of the longitudinal aquiferous canals, and the nervous cords; $m$, the homologue of the madreporiform body; $mc$, the madreporic canal; $h$, the heart; $o$, the aperture of $ov^3$, $ov^4$, the reproductive organ; $ms$, $ms^1$, $ms^2$, the pairs of longitudinal, muscular bands; $d$, the skin. — *Original*.

Fig. 116. An actual transverse section across the median pair of muscular bands (fig. 115, $ms$) of Caudina, and the nervous cord and aquiferous canal at that point; magnified 40 diameters. $m$, $m^1$, the edges of the muscular bands; $b$, a narrow, double, muscular band cut across; $f$, $f^1$, the edges of the circular muscular bands which lie immediately beneath the skin ($s$); $f^2$, $f^3$, a sort of fibrous tissue in which $f$, $f^1$, are imbedded; $s$, $s$, the skin, composed of numerous heterogeneously interwoven fibres; $c$, the aquiferous canal which lies in the furrow between $m$, $m^1$; $n$, the nervous cord lying just beneath the skin, external to the muscular layers, and imbedded in the fibrous tissue ($f^2$, $f^3$). — *Original*.

across, and the posterior end gradually tapers from a short distance behind the thickest mid-region of the body. As they lie on the beach after being cast up by a storm, the Caudinas, from their shape and fleshy color, might easily be mistaken, at a distance, for so many human fingers lying loose and idle. The mouth is a simple aperture in the middle of the truncate head. The latter is bordered right and left by a row of cross-tipped, hollow, cylindrical feelers, ($t$, $t^1$,) so disposed that one of them is in the lower middle line of the body, directly under the mouth, and a pair in the upper median line, and the remaining twelve lie, six on each side, at equal distances apart, between these two points.

From the mouth, the passage into the stomach is through a simple, broad, tubular throat, ($g$,) which expands slightly as it joins the latter. The stomach and intestine are so uniformly tubular that it is not possible to make a proper distinction between them, and I shall therefore designate the whole digestive canal as the intestine. This organ extends at first backwards, a little to the left of the median line, through two thirds the length of the body, and then abruptly making a fold, ($g^1$,) it advances along the right of the median line, as far as the anterior third of the body, and then again redoubling itself, ($g^2$,) it passes backwards with a gradual taper ($g^3$) to its posterior termination ($g^4$) at the end of the tail. Throughout its whole tract there is but one pair of hollow appendages, (figs. 114, 115, $rt$, $rt^1$, $rt^2$,) and these arise from that part of it which lies just behind its first fold, one from the left and the other from the right of its upper side. At their point of attachment ($rt$) they open widely into the intestine, in the same way as do the ten puckered appendages of the starfish (fig. 111, $st^1$). The main tube (figs. 114, 115, $rt^1$) of these appendages gives off in every direction numerous processes which branch like the twigs of a tree, and extend themselves throughout the whole circum-intestinal cavity, even to the extreme anterior end of the body (fig. 114, $rt^2$). On account of their peculiar conformation, and their supposed office as respiratory organs, they have been called the *respiratory branches*. This office is performed by the introduction of water through the posterior opening of the intestine into the tapering cloaca ($g^3$) and thence into the branches, where, by the enormous expansion of their branchlets, an extent of respiratory surface is bathed that is scarcely equalled in any other known group of animals.

The *nervous system* consists of a simple flat collar (fig. 114, *c*) which embraces the throat almost close up to the edge of the mouth, and five slender threads which extend along the sides of the body just beneath the thick skin (fig. 116, *s*). The position is indicated by these longitudinal rows of dashes, (fig. 114,) – – – – –, of which one trends along the lower mid-line ($aq^1$) of the body, and two along each side, right and left ($aq^2$, $aq^6$). In the foreshortened view these five nearly equidistant points (fig. 115, $aq^4$, $aq^3$, $aq^9$) indicate their position.

The *aquiferous circulatory system* is much more complicated than in the starfish. But in order that you may the better understand its distribution I must first state that the throat is encased in a sort of sheath, which consists of fifteen calcareous, stave-like bodies (fig. 114, *f*, *f*$^1$) joined side by side like the staves of a pail. They are usually designated as the *buccal plates*. Of these, ten, in pairs, at alternate intervals with the remaining five, extend backward beyond the latter, in the form of five broad forks (*f*). Their several positions right and left correspond with the five longitudinal nervous threads, and they serve as solid bases of attachment for five pairs of ribbon-shaped muscles, (fig. 115, *ms*, *ms*$^1$, *ms*$^2$, fig. 116, *m*, *m*$^1$,) which extend from just behind the base of the tentacles, along the sides of the body; one of each pair running on each side of the line along which the five nervous threads trend. The general lining of the body is a muscular layer (fig. 116, *f*, *f*$^1$,) which lies immediately within the outer skin (*s*), and consists of fibres which bind the body as with innumerable, circular, contractile hoops. Along the inner face of this layer the five pairs of longitudinal muscles (*m*, *m*$^1$) are attached, and along its outer face, next the skin, the five nervous threads (*n*) extend in so many lines, exactly over the several intervals which are embraced by the pairs of muscles. It will be seen presently that each of these intervals is occupied by one of the longitudinal canals ($c$) of the water circulatory system, to which we will now return. The canal which corresponds with the one which encircles (fig. 111, $aq$) the throat of the starfish, is suspended from the points of the forks (fig. 114, $f$) of the buccal plates, and pursues a rather irregular course ($r$) in its circuit. At each fork a canal passes from it directly forward to and into the five tentacles which lie in the same trend, and at the same time pours its contents into a circular passage ($aq^3$) which runs along the base of these tentacles, and thus indirectly supplies the ten alternating ones. From each of the five points where the canals from the forks enter the tentacles, a simple canal (figs. 114, 115, $aq^1$, $aq^2$, $aq^4$ to $aq^9$) arises which extends backwards along the side of the body, in the furrow that separates each pair of muscles (fig. 116, $m$, $m^1$) from one another. There are no external tubular feet like those of the starfish, and consequently no lateral branches from the longitudinal canals, except at their anterior points of origin, where they connect more or less directly with the cavities of the tentacles. The latter are all that remain, in this reducing process, to represent the numerous, tubular, external appendages of the Starfishes, Sea-urchins, and the lower ranks of Trepangs. The madreporic canal (figs. 114, 115, $m$ to $mc$) arises from the circular canal ($r$) in the superior middle line of the body, between the two upper forks of the buccal plates. It is bent upon itself several times so as to form a thick, convoluted, elongate mass, and is terminated ($m$) by the homologue of the madreporiform plate of the starfish, but, unlike that of the latter, is wholly included within the body cavity.

The *blood circulation* is carried on in a system of vessels which run along the upper and lower edges of the intestine, and connect with each other by multitudes of little cross branches. The latter, as they traverse the intervening space, are interlocked with the twigs of the respiratory branches, and in this way the blood is aërated. The principal trunk of this system is a thick tube ($h$, $h^1$) which courses along the back of the stomachal region of the intestine, and terminates (at $h^1$) close to, and in the same line with, the madreporic canal.

The reproductive organ consists of a thick tube ($o$ to $ov$) which at one end opens on the mid-line of the back, through a slightly raised protuberance ($o$) immediately over the madreporic canal, and at the opposite end forks into a dozen or fifteen tubules, (fig. 114, $ov^1$, $ov^2$, fig. 115, $ov^3$, $ov^4$,) that spread themselves right and left among the intestinal folds and the ramifications of the respiratory branches.

Mark the contrast now between these two foreshortened views; the Starfish (fig. 111) exemplifying the perfection of the type in a multiplied sameness, and the Trepang (fig. 115) setting forth the subordination of the type to the bilateral dominant, by the culmination of the reducing process which has established single organs or pairs of organs, at or on the right and left of a median line. The pair of respiratory appendages ($rt^1$) of the latter are multiplied in another form (fig. 111, $st^1$) five times in the former. The reduction of the rows of external

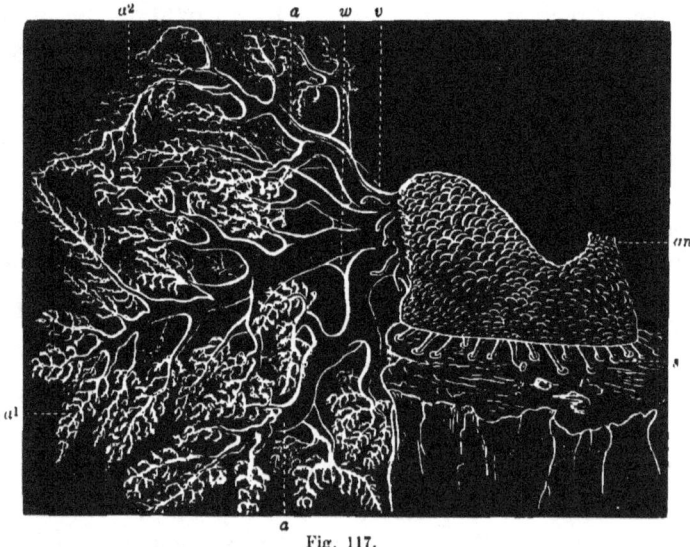

Fig. 117.

Fig. 117. *Psolus phantapus.* Strfldt. ⅓ natural size. From a specimen caught in Massachusetts Bay, and kept in the Aquarial Gardens, at Boston.

tubular appendages to nothing but the tentacles which crown the head progresses through a graduated series of changes, during which the upper two, on the right and left, are obliterated first. This is singularly exemplified in another of our native Trepangs, which is pictured in this diagram (fig. 117) in profile. Not only is it destitute of the two rows just mentioned, but the median one of the remaining three is nearly obsolete; thus leaving the lower right and left (*s*) rows to perform the office to which the Starfish devotes a fivefold number. But one step more, and the reduction leaves no traces — as in Caudina — of these appendages along the body. Coincidently with this process the transverse filaments which go to these feet from the nervous cords are also reduced, and thus the preponderance of this system is thrown to the side opposite the heart, — as these figures (p. 122, figs. 57, 58) of the Zoöphytic ideal type are intended to express.

This concentration of the nervous system, consequent upon the reduction of parts from a higher to a lower number, is everywhere manifest throughout the animal kingdom, and nowhere so conspicuously as among the jointed animals, the Articulata; and always coincidently with the elevation to successively higher grades of being. It is, therefore, not to be overlooked among the Zoöphytes, although it is exhibited in so feeble a manner as to almost escape detection unless upon the closest scrutiny.*

The body is covered by closely set, semicircular, calcareous scales. *a, a*, the branching tentacles; $a^1, a^2$, the finer ramifications of *a*, waving to and fro in the currents of water; *w*, the web between the bases of the tentacles, forming with the tentacles a deep, funnel-shaped vestibule to the mouth; *v*, vermiform appendages, appearing like an outer circle of feelers; *an*, the posterior opening of the intestine; *s*, the tubular appendages or suckers of the left, ventral row. — *Original.*

* In order to anticipate the objection which no doubt will be raised by the advocates of the radiate type, I would remark, first, that I am not oblivious of the apparent contradiction to this view in the process of the reduction of the ambulacral rows of Echini; and secondly, that it is only *apparent*, as may be ascertained by a seriatim inspection of the various groups of the order. The whole process culminates in producing a preponderance of the *antihæmal* pair; the so-

The blood circulating system, tenfold repeated in the Starfish, significantly approximates in Candina the singleness of character of that of marine worms, and occupies an almost identical relative position along the median line of the body. Lastly, but by no means less prominently, the reproductive system contrasts its uniformity with the tenfold repetitions of the homologous organs of the Starfish, and establishes itself, in conformity with the rule which the other organs have followed, in the strictest symmetrical relation to right and left, along the upper median line.

called odd ambulacrum being reduced first, and then the two on the hæmal side. In Holothurians the same result is arrived at by an inversion of the process as seen in the Echini; that is, the hæmal pair disappear first, next the single anti-hæmal one, and finally, carrying the process farther than in the Echini, the antihæmal pair vanish. The consequence in the latter case is the perfection of *nervous cephalization*, and therefore a ranking of the *apodous* Holothurians above the pedicellate ones.

## CHAPTER XI.

### MOLLUSCA.

There is not one of the five great groups of the animal kingdom in which, as in Mollusca, the relation of bilaterality to its subordinate, the type of the grand division, is so clearly lighted up, and yet at the same time in which the former seems to be so lost sight of and obliterated by the excessive development of the latter. We might well say that the Mollusca are characterized by the *excessive development of uniformity*. Among Zoöphytes and Articulata we find an excess of repetitions, of multiplicity of parts; but here every part or organ is *single*, or a single pair, and, instead of repeating itself, it develops its uniformity to relatively extreme proportions. The Zoöphytes are, from back to front, dorso-ventrally, polymerous; the Articulata are, from tail to head, uro-cephally, polymerous; and the Mollusca are *monomerous*.

An attentive inspection of these two, (figs. 118, 119,) profile and foreshortened, views of an animal of the *lowest* group among Mollusca will instruct you as to how plainly bilateral even the least developed of this grand division are; and now transporting your eyes to another pair of figures, (figs. 124, 125,) which represent a profile and foreshortened view of one of the *highest* and most completely organized of Mollusca, you will find the same idea none the less manifest. Among the intermediate grades of these two extremes the apparent obliteration of this character is most noticeable; and in order to unravel before your eyes the obscurity which reigns there, it is necessary to make you acquainted with the relations and nature of the organs of the more simple members of this great group. I shall return, therefore, to the lowest ranks, and make a brief sketch of the Bryozoan organization.

*Pectinatella.* Imagine a cylindrical tube to one end of which another tube, bent in the form of a horseshoe, (figs. 118, 119, *lp,*)

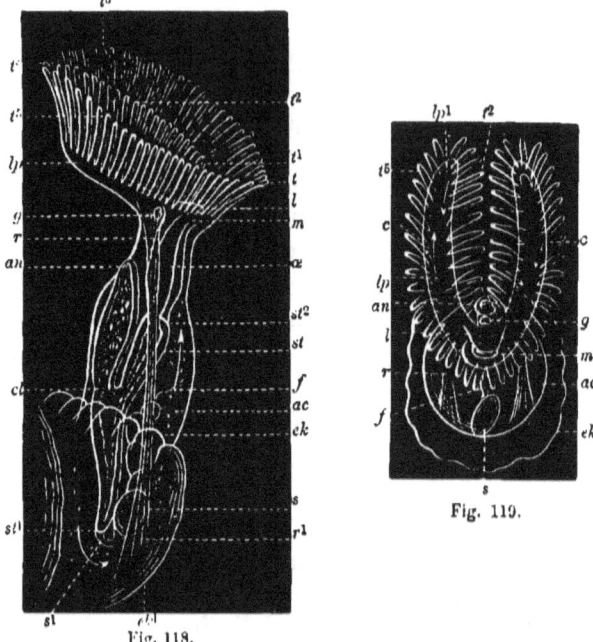

Fig. 118.

Fig. 119.

is attached at a point which is equally distant from its two ends,

Fig. 118. *Pectinatella magnifica.* Leidy. 20 diam. A profile view of one of the individuals of a compound, fresh-water Bryozoan. *ek, ek¹,* the gelatinous envelope of the colony; *t* to *t⁵,* the U-shaped double row of feelers; *lp,* the U-shaped arms; *m,* the mouth; *l,* the lip of *m; œ,* the throat; *st,* the stomach; *st¹,* the bottom of *st; st²,* the anterior end of *st; cl,* the valvular passage from *st* to the last division (*cl* to *an*) of the intestine; *an,* the posterior opening of the intestine; *ac,* the abdominal cavity, in which the arrows indicate the circulating currents; *g,* the nervous ganglion; *r* to *r¹,* the retractor muscles of the head; *s,* the *statoblasts,* or so-called winter-eggs, in various stages of development from *f* to *s; f,* the posterior end of the *funiculus,* attached to the side of the body; *s¹,* the anterior attachment of the reproductive organ. — *Original.*

Fig. 119. The same as fig. 118. A foreshortened view of the head, with the body in the distance. The letters as in fig. 118, and in addition, *c,* the membrane which passes from the base of one feeler to the other; the arrows indicate the direction of the currents of the circulating fluid in the U-shaped arms; *lp¹,* end of *lp.* — *Original.*

that is, at the toe of the shoe, and from the convex and concave curves of the U a row of hollow fringes ($t$ to $t^5$) projecting about in the same direction as the first tube, and you have the outlines of the general cavity of this animal. All three of these parts are in open communication with each other; and it is here only that the fluids, or rather fluid, of the body circulates. There is nothing that otherwise resembles a circulatory system. If, now, the U be divided right and left into halves, the line which separates them may be projected in the form of an imaginary plane so as to divide also the first tube, which is the body proper, into halves. This we will designate as the *axial plane;* and you will find that the arrangement of the organization is in reference to this plane.

Going back to the U, now, there is to be seen between the outer ($t^3$, $t^5$) and inner ($t^2$) rows of fringes, and immediately opposite the toe of the horse-shoe, a round aperture ($m$) half covered by a sort of lid ($l$). This is the *mouth*. From it the intestinal canal projects into the depths of the body cavity, and then doubles upon itself, and terminates ($an$) not far from the beginning, without deviating either to right or left from the axial plane. A knife passed along this plane would split the digestive system, from end to end, into halves. The throat ($œ$) extends about one fourth of the length of the body, and is shut off from the stomach ($st$) proper, except during the passage of food, by a sort of valve ($st^2$) which projects like a ring between them. The stomach ($st, st^1$) occupies the next two fourths of the general cavity, and preserves the same narrow proportions which obtain in the throat; *i. e.*, it is a mere thick-walled tube. Instead, however, of being continued directly from the furthermost point of its backward reach, it opens at its mid-length (at $cl$) into a thick tube, which passes forward and debouches ($an$) at the surface of the body, on the side next the two limbs of the U, and about opposite to the mid-length of the throat ($œ$).

All that has ever been discovered of the *nervous system* of Bryozoans is a small, double-oval, or broad, heart-shaped mass, ($g$,) which trends transversely to the bilateral plane, and is situated close to the anterior end of the throat, on the side facing toward the limbs of the U, and therefore in the interspace between the anterior (*m*) and posterior (*an*) ends of the digestive canal. From this there are a few nervous threads which pass, in part, right and left into the fringed arms, and otherwise into the throat near the mouth.

The *reproductive system* (*s, f*) is attached to the wall of the general cavity, on the side opposite to the nervous centre, (*g,*) and consists of a more or less elongated mass within which the eggs of various sizes are imbedded.

The animal has the power of withdrawing into itself by what is called invagination; like the sliding of the tubes of a telescope one within the other, or more properly, like the inversion of the finger of a glove upon itself. This is accomplished by the aid of two long, complex muscles (*r*, *r*$^1$) which extend, one on each side, from the head to the bottom of the sheath (*ek*, *ek*$^1$) which encloses the organization. The sheath of the species before us is peculiar on account of its jelly-like character; and as the individuals live in a community, which has arisen by the budding of one from the other, their combined sheaths form a large tremulous mass, oftentimes six inches in width, covering submerged sticks and the stems of aquatic plants in slow streams and pools.

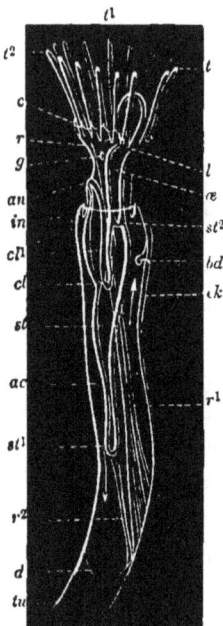

Fig. 120.

The sheath of most of the Bryozoa is more elongated than the one we have before us, and usually of a parchmenty, horny, or coralline nature. This other figure (fig. 120) represents one, *Fredericella*, which possesses a parchmenty sheath, (*ek*, *tu,*) and whose crown of tentacles (*t, t*$^1$, *t*$^2$) is a nearly perfect circle, but still shows a trace of that prepon-

Fig. 120. *Fredericella regina.* Leidy, MSS. 25 diam. A profile view of one of the individuals of a branching, compound, fresh-water Bryozoan. *tu*, the parch-

derance toward the nervous (neural) side (*g*) which is so largely developed in the one which I have just sketched out. So slight indeed is this obliquity, from front to back, of the tentacular crown of Fredericella, that it seems almost like an accidental one-sidedness; but an attentive examination of any number of individuals will reveal the unvariableness of this character, and always in symmetrical relation to right and left; although not so strikingly prominent in this respect as in Pectinatella.

Among the asymmetrical forms of Mollusca, and one of the most lowly organized of the series, is the familiar *Oyster*. Although so comparatively simple in its structure, there are numerous intermediate stages of organization between it and the Bryozoans, some of which, from their peculiar interest, I would be glad to lay before you; but I must, from want of time and space, content myself with the illustration of a sketch here and there, along the line of upward progress, of such as are most serviceable for our immediate wants. The irregular, ragged outline, and unsymmetrical shape of the Oyster (fig. 121), would hardly lead one to suspect the unity of relationship to right and left which reigns in its organization. That it has a definite right and left, and that too always corresponding, the first to the flat, and the latter to the deeper hollowed shell, is easily demonstrated, and without entering into the minuter details of its structure. Having separated the flat valve from the animal, keep it in position with the right hand, and hold the hollowed valve ($sh^1$) in the left hand, both so disposed that the thicker part or beak (*sh*) of the shells is turned from the body and their edges project in a

menty exterior tube common to the whole colony; *ek*, the thinner transparent part of *tu* immediately about the body of the contained individual; *t*, $t^1$, $t^2$, the slightly oblique single circle of feelers; *c*, the *calyx*, a scalloped membrane which joins the bases of the feelers; *l*, the lip at the front of the mouth; *œ*, the throat; *st*, the stomach; $st^1$, the bottom of *st*; $st^2$, the anterior end of *st*; *cl*, the valve between *st* and $cl^1$; $cl^1$, the cloaca or last division of the intestine; *an*, the posterior opening of the intestine; *ac*, the abdominal cavity; *g*, the nervous ganglion; *r*, $r^1$, the retractor muscles of the head; $r^2$, the retractor muscle of the stomach; *bd*, a very young bud; *in*, the walls of the body partially doubled upon themselves, *invaginated*; *d*, the posterior end of the individual. — *Original*.

perpendicular plane, and you will have the head (*m*) of the animal turned from you, its tail nearest to you, and its back upwards. Consequently, the flat valve is the right one and the hollow valve the left.

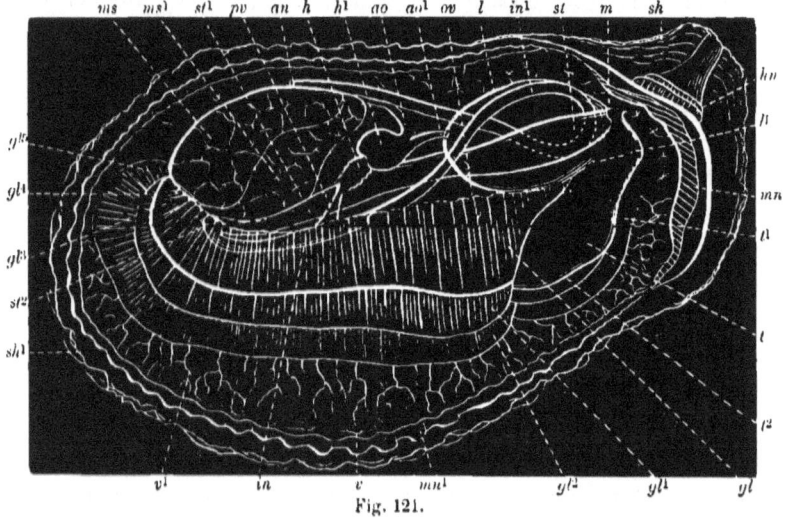

Fig. 121.

Taking off, now, the right valve, and letting the other one lie over a little in your left hand, but in the same relative position

Fig. 121. *Ostrea Virginica.* L. Natural size. The right mantle of the Oyster removed, and the internal organization displayed. *sh*, the anterior, and *sh*¹ the posterior end of the shell; *hn*, the ligament which acts as a hinge to the shells; *mn*, the line along which the right mantle was cut away from the left; *mn*¹, the edge of the left mantle; *m*, the mouth; *st*, stomach; *st*¹, posterior end of *st*; *st*², first bend of the intestine (*in*); *in*¹, the intestine where it is buried in the liver (*l, l*¹); *an*, posterior end of the intestine; *l, l*¹, the upper and lower sides of the liver; *t*, the right, and *t*¹ the left halves of the outer, leaf-like appendage of the mouth; *t*², the edge of the inner, leaf-like appendage; *gl*, the anterior, and *gl*³, the posterior ends of the right and outer gill; *gl*¹, *gl*⁴, the inner gill of the right side; *gl*², *gl*⁵, the inner gill of the left side; *h*, the auricle, and *h*¹ the ventricle of the heart; *ao*, the posterior, and *ao*¹, the anterior aortas, or distributing branches of the circulatory system; *pv*, the branchio-cardiac vessels returning the blood from the gills to the heart; *v, v*¹, blood-vessels in the mantle; *ov*, the position of the reproductive organ, just behind the liver; *ms, ms*¹, the two halves of the adductor muscle. — *Original.*

as before, lift up the thin, fringed, veil-like covering, (the mantle,) and you expose the *gills*, ($gl$ to $gl^5$,) which hang — as represented in this figure — from the lower side of the animal. They are four in number, and have the appearance of cross-ribbed bands which stretch from one end of the body to the other. It is in their inequality that the one-sidedness of the animal is most prominently set forth. The two ($gl$, $gl^3$, and $gl^1$, $gl^4$) on the right side of the body are of unequal width, and both are narrower than the left pair ($gl^2$, $gl^5$); the latter extending much nearer to the edge of the mantle than the former. In front of the gills are four smooth, leaf-like bodies, ($t$, $t^1$, $t^2$,) which hang in pairs below and right and left of the mouth, ($m$,) and, being joined above, form a sort of hood over it. They are generally considered as organs of touch or prehension, and, like the gills, are of unequal width; the narrower ($t$) pendent on the right, and the larger ($t^1$) and longer on the left.

Near the posterior end of the gills, and above them, is the great double muscle ($ms$, $ms^1$) which serves to keep the valves closed. In the shells its position may be recognized by the chestnut-colored, striped, agate-like spot, as big as one's thumb-nail.

Immediately in front of the muscle is a hollow, which, in freshly opened specimens, seems to be the theatre of very active operations. On carefully removing the semi-transparent mantle with a pair of sharp scissors, the cause of this phenomenon becomes apparent. We find there a sort of double sac, ($h$, $h^1$,) the two halves of which are constantly and alternately contracting and expanding, in moderately rapid succession. This sac is the *heart*. The anterior half, the *ventricle*, ($h^1$,) by contraction forces the blood through the two vessels ($ao$, $ao^1$) which go off from its narrower end into the body, and thence into the minute vessels ($v$, $v^1$) which branch through the mantle ($mn$, $mn^1$). From these, by means of other minute currents, the circulating fluid is brought back into the gills, ($gl^1$ to $gl^5$,) where aërification takes place, and then the blood is returned, through the so-called branchio-cardiac vessels, ($pv$,) to the posterior chamber, *auricle*, ($h$,) of the heart. From the latter the blood is injected into the anterior chamber, ($h^1$,) and the circulation again repeats the round from the ventricle ($h^1$) into the body and mantle, thence to the gills, and finally back to the auricle.

The *digestive system* consists of a tube, of varying thickness, which doubles upon itself twice in its course from mouth ($m$) to vent ($an$). From the mouth ($m$) the stomach ($st$, $st^1$) broadens backwards, without the intervention of a throat, in the mass of brown liver ($l$, $l^1$) which surrounds it, and then slightly narrowing, it passes in a direct line, below the heart and along the lower side of the muscle, ($ms$,) almost to the posterior end of the gills. There it abruptly narrows into an intestine, ($in$,) which, making a sudden bend, ($st^2$,) passes pretty closely along the right side of the stomach, back to the liver, ($l$, $l^1$,) into which it enters and passes along near its right side, (at $in^1$,) and then across its forward end, nearly over the mouth, ($m$,) and down the left side, and then, making a long curve, rises again toward the upper side of the liver, and, passing out of it, bridges over the space in which the heart lies, and comes to a termination ($an$) on the upper side of the adductor muscle, ($ms$,) and to the left of the line along which the stomach passes.

The nervous system consists principally of two widely distant masses or groups of ganglions. One of them lies across the upper lip of the mouth, and the other below the great muscle near the first bend of the intestine; and the two are connected with each other by delicate nervous threads which pass along on each side of the body. The anterior mass is called the " cerebral ganglion," or *brain*, and the posterior one the " branchial ganglion," on account of its supplying large nerves to the gills, as well as to other neighboring parts.

The reproductive organ ($ov$) lies just behind the liver and under the stomach. It is an irregular mass, which, with the liver and certain other smaller bodies of an undetermined nature, forms a basis on which the stomach and intestine lie, or pass through.

These organs, and the great double muscle, form the bulk of the body, upon which the right and left mantle is laid, and from which the four gills hang in long, parallel, unequal strips. Everything is one-sided, and as if the body had been trodden upon and flattened out unequally, in such a way as to force the whole right half up toward the back.

In some kinds of oysters, e. g., Gryphæa, the left valve is so deep, and so much curved upon itself, in fact, partially rolled up spirally, as to leave but little of the body to be covered by the right valve. In this respect they approximate the form of the shell of the Snail, and the included body is likewise modified so as to conform to this configuration.

In the *Snail*, (Helix,) the one-sided development of the body is carried to the highest pitch of asymmetry that is observable among Mollusca. It commences its career with a perfectly

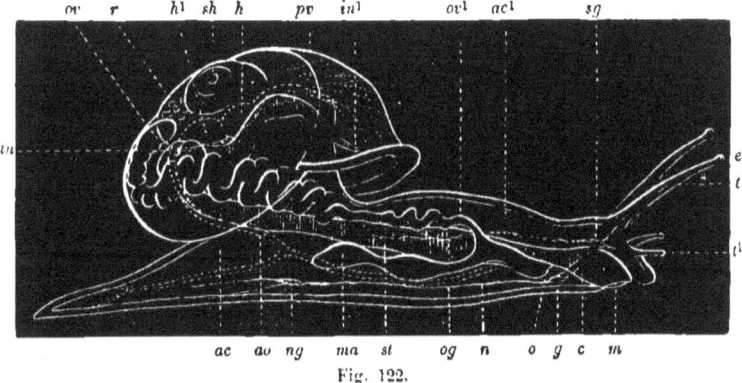

Fig. 122.

symmetrical body, but ere long the right side outgrows the other, and finally the whole organism has the spiral conformation represented here (figs. 122, 123). All that is usually seen of a

Fig. 122. *Helix albolabris.* Say. Diagramic representation of the common Snail. 2 diam. $ac, ac^1$, the abdominal cavity ; $sh$, the shell ; $t$, the larger pair of feelers, with an eye ($e$) at the tip of each; $t^1$, the smaller pair of feelers; $m$, mouth ; $st$, stomach ; $in$, intestine ; $in^1$, posterior opening of $in$ ; $sg$, the superior ganglion of the head ; $g$, the inferior, or sub-œsophageal ganglion ; $c$, the nervous collar ; $n, ng$, the foot nerves ; $og$, the œsophageal, or gullet nerves ; $h$, the auricle, and $h^1$ the ventricle of the heart ; $ao$, the aorta, or main artery ; $pv$, vein from the lung, or pulmonic vein ; $ov$, the ovary, or egg-bearing organ ; $ov^1$, the oviduct, or emptying conduit of $ov$ ; $o$, exterior aperture of $ov^1$ ; $r$, the fertilizing gland, or male element of the reproductive organs ; $ma$, the matrix. — *Original*.

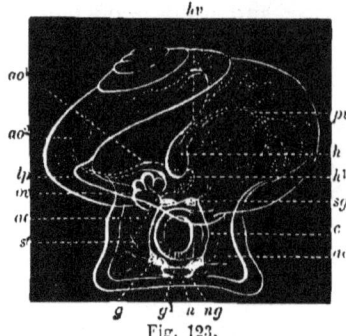

Fig. 123.

Snail is the forepart of its body, including its head, ($t$, $t^1$,) the muscular disc, or foot ($m$ to $ac$) upon which it creeps, and that portion which lies immediately above the latter and includes the stomach, ($st$,) the bulk of the nervous system, ($sg$, $c$, $g$, $n$, $ng$,) and the emptying conduits of the reproductive organ ($or^1$, $o$). Within the shell are included the posterior half of the digestive system, the reproductive organ, ($ov$,) the heart, ($h$, $h^1$,) and the lung ($pv$). That part of the body which is usually extended from the shell ($sh$) has an elongated spindle shape, and is flattened upon the lower side so as to form the creeping disc. At one end, which is the head, it is truncated, i. e., as if cut straight across. The head has four feelers, of which the upper pair ($t$) are the longer, and have each an eye-spot ($c$) at the end. The lower pair ($t^1$) are mere tactile organs.

The *mouth* ($m$) is situated on the lower side of the body, and just behind the truncate end. From it the stomach ($st$) extends directly backwards for about two thirds the length of the body, and then narrowing into an intestine, with varying convolutions, passes within the spire of the shell, ($sh$,) and bending upward, forward, and considerably to the right, comes to a termination ($in^1$) near the upper right edge of the aperture of the latter.

The *circulatory system* consists of a heart ($h$, $h^1$) with two cavities, one for receiving and the other for the distribution of

Fig. 123. A foreshortened, diagramic view of fig. 122. *lp*, the lip at the anterior border of the mouth of the shell; *st*, the stomach; *ac*, the abdominal cavity; *h*, the auricle of the heart; $h^1$, the ventricle of the heart; *hv*, the vein which brings the blood from behind to *h*; *pv*, the pulmonic vein; *ao*, the aorta coming forward; $ao^1$, the aorta which branches at $ao^2$, in the region of the liver and reproductive organs; *sg*, the super-œsophageal nervous ganglions; *g*, the sub-œsophageal ganglious; $g^1$, the inferior edge of the nervous collar; *c*, the lateral threads of the nervous collar; *ng*, the nerve which branches in the foot; *n*, transverse branch from *ng*; *ov*, the reproductive organ. — *Original*.

blood, and the vessels which carry the circulating fluid through the body. The heart lies obliquely across the mid-line of the body, considerably to the left of the forward bend of the intestine, and close to the surface of the back. From one cavity of the heart, *i. e.*, the ventricle, ($h^1$,) the blood is expelled into the arteries, (*ao, ao$^1$, ao$^2$*,) which branch through all parts of the body. From the minute tips of the arteries the fluid then passes into return channels, the veins, one part of which unite in a vessel (*hv*) which goes direct to the auricle, (*h*,) and another set of them carries the blood to the net-work of vessels which branch through the lung. After being aërated, the blood passes from the smaller vessels into a single larger one, (*pv*,) which empties into the receiving chamber, the auricle, (*h*,) of the heart. By the contractions of the latter the blood is thrown into the ventricle, ($h^1$,) and thence goes out, and circulates as I have just described.

The *nervous system* is concentrated chiefly about the head, but sends off branches to various parts of the body. It consists of a double ganglion, (*sg*,) the brain so-called, which lies just above and across the throat, and of a still larger and broader double nervous mass (*g, g$^1$*) which rests beneath the throat, and transverse to the axis of the body. The upper and lower pair are connected with each other by double or triple nervous threads, (*c*,) which form a collar as they pass from above downwards and backwards on each side of the throat. From the upper pair nerves pass forward to the feelers, (*t, t$^1$*,) and backwards (*og*), on the right and left, over the stomach (*st*). The great nervous trunks (*n, ng*) originate one from each of the lower pair (*g*) of ganglions in the head, and pass backwards along the sides of the body, in nearly unbroken continuity, and with gradually diminishing diameter, until they vanish in the skin and muscles of the creeping disc. Where they branch there is more or less of a thickening, (*ng*,) but scarcely deserving the name of a ganglion.

The *reproductive system* (figs. 122, 123, *ov, ov$^1$, ma, o*) lies principally on the right side. Its essential part, (*ov*,) that in which the eggs are developed, is deep within the spiral portion of the shell, and the oviduct, ($ov^1$,) i. e., the canal through which the eggs pass to the outer world, extends in variously convoluted folds to a point near the right side of the head, where it opens (*o*) to give egress to its contents. Not far behind the outlet of the oviduct, a tubular blind-sac (*ma*) opens into the latter. It serves as a reservoir for the fertilizing fluid which is poured over the eggs as they pass along the oviduct ($ov^1$) toward its aperture.

If now we compare the relative position of these organs with that in the Oyster, (fig. 121,) we shall find that whereas in the latter the bulk of the organization is, as it were, pushed over to the left side, in the Helix it is reversed in this respect, and is enclosed entirely in a deep spiral shell, which corresponds to the right valve of the oyster. There is no left valve in the Helix, but in other Gasteropods it exists in the form of a horny, or shelly, flat, spirally marked shell, which is attached to the back of the creeping disc, near its posterior end. When the animal withdraws into its shell, this valve, the *operculum*, fits the aperture closely, and protects the body from all intruders. The most noteworthy features in Gasteropoda, as represented by Helix, are the head-like configuration of the anterior end, and the great preponderance of the nervous ganglia in that region. These characteristics we find carried to a much higher degree in the Cephalopoda, the highest order of Mollusca.

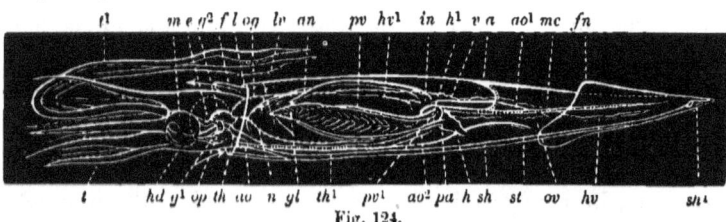

Fig. 124.

Fig. 124. *Loligopsis illecebrosa.* Les. ⅔ of natural size. A profile view of the left side of a Squid. *t*, the eight shorter arms: $t^1$, the pair of longer arms; *hd*, the head; *fn*, the fin; *f*, the funnel; *l*, the edge of the mantle; *mc*, the cavity of the mantle; *sh*, the shell; $sh^1$, the conical hollow of *sh*; *m*, the jaws; *th*, $th^1$, the throat; *st*, the first stomach or *crop*; *in*, the intestine; *an*, the posterior end of the intestine; *lv*, the liver; *h*, the auricle of the heart; $h^1$, the ventricle of the heart; *ao*, $ao^2$, the aorta which carries the blood forward; $ao^1$,

*Cephalopoda.* As a representative of this order, I have taken the common Squid, (figs. 124, 125,) a species of Cuttlefish that abounds along our sea-shores during the warmer months. It is represented here in a reversed position from that in which it swims, in order to bring the organs into the same relation, in regard to up and down, as those in the figures of the Snail and the Oyster have.

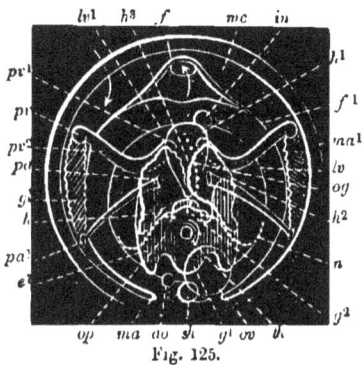

Fig. 125.

It usually swims with its back downwards, as I have already stated, (p. 121, note,) and most frequently backwards; although it moves with equal facility head foremost. In order to understand its mode of swimming you have to learn that its body is enclosed up to the neck in a loose muscular sac, (fig. 125, $ma^1$,) to

the posterior aorta; $a$, a branch from $ao^1$ going to the mantle; $lv$, the posterior, and $lv^1$ the anterior veins, which unite in a common reservoir at $v$, from whence the blood enters the auricle ($h$); $pa$, the point of origin of the artery which passes along the lower side of the gill ($gl$); $pv$, the branchial vein; $pv^1$, point of entrance of the blood from the gill, through $pv$, into the ventricle ($h^1$); $gl$, the left gill; $g^2$, the super-œsophageal nervous mass; $g^1$, the sub-œsophageal nervous mass, or lower half of the nervous collar; $op$, the optic nerve, cut across; $og$, the superior, visceral nerve; $n$, the mantle nerve; $e$, the pupil of the eye; the outer, dotted circle indicates the outline of the eye; $ov$, the reproductive organ. — *Original.*

Fig. 125. The same as fig. 124, representing a diagramic view, from the posterior end. $ma^1$, the mantle; $f$, the funnel; $ma$, inferior edge of the membranous, valvular prolongations ($f^1$) from the funnel ($f$); $mc$, cavity of the mantle; $sh$, the shell; $th$, the throat; $in$, the mass which includes the posterior portion of the intestine, the oviduct, and the *ink-bag;* $lv$, $lv^1$, the liver; $h^2$, $h^3$, the right auricle; $h$, the left auricle; $h^1$, the ventricle; $ao$, the anterior aorta; $pa$, $pa^1$, the artery going to the gills ($gl$); $pv$, $pv^1$, $pv^2$, the branchial vein, emptying at $pv^1$ into the ventricle ($h^1$) of the heart; $gl$, the gills, cut across; $g^2$, the super-œsophageal, and $g^1$, the sub-œsophageal, nervous mass; $op$, the optic nerve; $e^1$, the expansion of $op$ at the back of the eye; $og$, the superior, visceral nerve; $n$, the mantle nerves; ($og$ and $n$ are dotted lines;) $ov$, the reproductive organ. — *Original.*

which it is attached along the lower side, (*ma*,) from head to tail. On the upper side of the neck there is a hollow, conical projection (*f*) called the *funnel*, from the base of which a flap (*f*¹) passes down each side of the body to the point (*ma*) where the latter is joined to the sac (*ma*¹). By the help of these two parts, then, the muscular sac, *i. e.*, *mantle*, and the *funnel*, all the various motions of the body are accomplished. When the edge (fig. 124, *l*) of the mantle is open at the neck, water flows in until the space (*mc*) around the body is filled. If, now, the animal wishes to swim, it contracts the mantle, and the water consequently seeks an outlet; but as the flap (*f*¹) of the funnel acts as a valve by pressing against the inner face of the mantle, the water is prevented from going out by the way it came in, and therefore is projected with great force through the funnel (*f*) toward the head, and the reaction of the outgoing current propels the body in the contrary direction, *i. e.*, backwards. In order to reverse the direction the funnel is bent upon itself, as you would flex the finger, and then the water being forced out toward the tail, the reaction moves the body head foremost. If the creature wishes to turn round, the funnel is simply bent to one side, and the reaction of the excurrent water throws the head in the contrary direction. At the posterior end of the body there is a fin-like organ, (*fn*,) which is attached along its mid-line to the lower side of the tail. When the Squid is swimming, this fin is usually folded around the tail; but whilst turning or moving gently, it is used as a balancer, by spreading it out, and waving its edges up and down, as fishes do with the fins on each side of the neck.

The *head* (*hd*) is set off from the rest of the body by a slight constriction, or neck. Around the mouth, (*m*,) which is at the end of the head, there are arranged, right and left, ten arms, (*t*, *t*¹,) five on one side and five on the other, set so closely together as to form a complete circle above, at the sides, and below. Of these ten, eight (*t*) are in one circle, and the other two (*t*¹), much larger than the rest, are attached one on each side immediately within the circle of eight, and in the interval between the first and second upper arms of the right and left sides. The whole inner face of each of the eight shorter arms is covered from base to tapering tip with a double row of suckers, by which the animal adheres with great tenacity to its prey, and in fact to the hand that makes it captive. The longer pair of arms ($t^1$) have a uniform thickness from the base to near the tip, where they expand moderately into a sort of spindle-shaped disc, covered by numerous suckers like those on the shorter arms.

With these arms, and a pair of large, staring eyes, ($e$,) the Squid may be truly said to have a formidable aspect. And such it proves itself to be to one who may incautiously take hold of it; quick as a cat it throws its head around to the hand which seizes it, fastens its slimy arms to the skin, and buries its sharp, hooked jaws ($m$) in the flesh. Fortunately, our native species is but just large enough to draw blood from the hand. Rarely does the body exceed ten inches in length. The foreign species, some of which have arms as thick as a man's thigh, and jaws as large as those of a snapping-turtle, are much more formidable.

Most of the internal organization is included under the cover of the mantle, the organs of the head alone being excepted.

The *mouth* is a highly distensible aperture which lies at the bottom of a sort of cup which is formed by the circle of arms and a membrane which extends between the bases of the latter. Immediately within the mouth is a pair of horny jaws, ($m_1$,) placed one above and the other below, in such a way that the upper slides over the lower one so as to cut like a pair of shears. From this point the throat, forming a narrow tube, ($th$,) passes obliquely through the neck to the lower side of the body, where it extends ($th^1$) along the middle line, with a moderate increase in thickness, to a point just behind the first half of the body, and immediately under the heart ($h$). At this place it expands sideways, to the left, into the first stomach, ($st$,) or *crop;* a highly muscular, oval organ, which tapers away posteriorly and ends about half-way to the end of the tail. After giving off this blind sac, it becomes the intestine, which proceeds but a short distance, and then opens into the true stomach. This is a large sac which lies on the right side of the inferior mid-line, and just behind the heart. From this the intestine bends forward and upward, and, passing under the right limb of the heart, rises (at $in$) to the upper surface of the liver ($lv$, $lv^1$), and terminates ($an$) not far behind the neck, slightly to the left of the middle dorsal line, and near the base of the funnel ($f$).

The liver ($lv$, $lv^1$) is an immense, elongate oval, brown mass, which extends from the neck to the heart, and occupies nearly the whole space between the upper and lower, and right and left sides of the visceral mass in this half of the body.

The *circulatory system* consists of a triple chambered heart, and arteries and veins. The *heart* ($h$, $h^1$, $h^2$, $h^3$) lies at the posterior end of the liver, ($lv$,) and exactly in the middle line of the body. The main chamber, the *ventricle*, ($h^1$,) is more elevated than the two *auricles*, ($h$, $h^2$, $h^3$,) on its right and left. In the process of circulation, the blood issues from the ventricle and passes into the great anterior ($ao^2$, $ao$) and posterior ($ao^1$) arteries, the first running along the lower side of the body close to the left of the throat and into the head, and the second extending along the upper face of the posterior visceral mass, and thence, after dividing into numerous large and small branches, to the various organs. From the tips of the various arterial branches the blood escapes into return channels, and these coalescing into larger vessels, the anterior ($hv^1$) and posterior ($hv$) veins, the circulating fluid is emptied into the right ($h^2$, $h^3$) and left ($h$) auricles. From these the blood is injected into the right and left gills ($g?$) through a vessel ($pa$, $pa^1$) which runs along the lower edge of each, and branches within them in regular, parallel, transverse channels. The blood, being thus aërated, is taken up by the return currents in equally minute, parallel channels, and poured into a larger vessel, ($pv$, $pv^1$, $pv^2$,) which, passing along the upper side of each gill, empties its contents into the right and left (at $pv^1$) sides of the ventricle; and thus the circuit is finally completed.

The *gills*, ($g?$,) which perform so important a part in this system, are two elongate, leaf-shaped bodies, placed symmetrically, one on the right and one on the left of the body, half-way between the upper and lower sides, and extend from the heart nearly to the anterior end of the liver. They are free from the latter, but are attached along one edge to the mantle; and thus, during the contractions and expansions of the latter, the water is brought in large quantities against the surface of these respiratory organs, and the commingled air is taken up and absorbed by the blood within their numerous capillary vessels. This process is called the *aërification* of the blood, and corresponds to the breathing of air in the lungs of warm-blooded animals.

The *nervous system* preponderates largely in the head; and we find here, as in all of the higher groups of animals, that the tendency toward the head, *cephalization*, is a marked feature in the organism. The great centre of this system may be likened to a thick, broad, heavy ring, (fig. 125, $g^1$, $g^2$, *op*,) lying immediately behind the jaws, and through which the throat, (*th*,) or gullet, passes. It is situated pretty close to the lower side of the head. Its principal regions are four in number, namely, 1, the so-called *optic lobe*, ($g^1$,) which is a prominent bulging on the lower middle line of the main mass; 2, a pair of small lobes, one on each side of, but slightly posterior to, the third; 3, another prominent swelling ($g^2$) on the upper median line; and 4, the *optic ganglions*, ($e^1$,) two swellings, or rather expansions, one on each side and between the first and third masses, and so wide in their extent that they cannot be said to lie either above or below, but rather on the equatorial line. It is equally clear, too, that the optic nerve (fig. 125, *op*) for either eye springs from a point (fig. 124, *op*) exactly opposite the side of the throat (*th*). From the foregoing it would seem to be pretty evident that the upper ($g^2$) and lower ($g^1$) median lobes must share equally in the duty of forming by their junction the ganglions which supply the organs of vision, and that the so-called optic lobes, the first, are not exclusively devoted to the eyes. The principal nerves which originate from this centre are few, but easily demonstrable. In the fore part of the head there are two sets: one above, known as the throat, or *œsophageal* nerve and ganglion, and one below the gullet, called the *buccal* nerves and ganglion, because they supply the parts about the mouth. From behind the brain ($g^2$) three great nerves originate, namely, a single median and two lateral. The single median one (*og*) arises from the upper median lobe, and passes along the upper side of the liver, (*lv*,) and branches to supply the various organs of the viscera. On this account it is called the *visceral nerve*. The two lateral nerves (*n*) originate from the pair of small lobes which lie on each side of the great upper one, ($g^2$,) and belong to a system which has such complicated relations as to puzzle comparative anatomists in regard to what they correspond with in the group to which the Snail belongs. They are termed the mantle nerves; but as they are also continued, on each side, to the extreme posterior end of the body, and there supply the highly muscular fin, (*fn*,) in all probability they correspond to the great foot-nerves of the Snail (fig. 122, *n*, *ng*); and consequently the two small lobes from which they arise, notwithstanding they are situated rather above the level of the throat, homologize with the widely separated pair of ganglions, *sub- œsohageal*, (figs. 122, 123, $g$,) which lie at the lower borders of the gullet of the Snail.

The *reproductive organ* (figs. 124, 125, *ov*) is an oval or spheroidal body, varying in size according to the season of the year, which lies to the right of the first stomach (*st*). Its outlet, *oviduct*, is a narrow canal which passes upward and forward, and opens near the posterior end (*an*) of the intestine.

The ink-bag lies close to the side of the posterior end of the intestine, (at $hv^1$,) and opens near the aperture of the latter. When alarmed, the animal beclouds the water about it by expelling the dark brown contents of this bag.

The rudiment of a *shell* (*sh*, $sh^1$) is imbedded in the thickness of the mantle along the whole length of its lower middle line. It consists of a semi-transparent, amber-colored, delicate, hornlike substance, and has the shape of a straight sword-blade, gradually widening toward the point; and at its posterior end is fashioned into a hollow cone ($sh^1$) with the concavity facing forwards and obliquely upwards. In position it corresponds with that of Spirula, another ten-armed cephalopod, which has a highly organized, cross-partitioned, spiral shell, very much like that of the common Nautilus, but placed in exactly the reverse position from the latter. The shell of the Nautilus occupies the same relative position as that of the Snail, whilst the shell of the Cuttle-fish, Squid, Spirula, &c., is placed upon the opposite side, i. e., the lower side of the body.*

* Looking at Nautilus from this point of view, the highly muscular "hood" falls very naturally into the place of a *creeping disc*, or foot; and in this respect it agrees in position with what the old Hollandish naturalist Rumphius ascribes to it when the animal is creeping with the shell above the body, like that of a Snail.

## CHAPTER XII.

### ARTICULATA.

THERE is one feature in the organization of Articulata, and indeed the most prominent one, which would appear to be strongly demonstrative that this grand division ought to be classed in a lower rank than that of Mollusca. I refer to the repetition of similar parts in Articulata, as contrasted with the total absence of this character in Mollusca. Were it confined to the exterior, we might ascribe to it a merely functional character, a jointing of the shelly or parchmenty covering in order to produce a complete flexibility of the body; but, as we find it equally conspicuous among the internal organs, the intestines, heart, lungs or gills, nervous system, and the reproductive system, one cannot avoid the conclusion that it is *typical* of the whole organization.

As it has already been shown (pp. 82, 84) that in many cases this repetition of parts is merely the multiplication of a kind of obscure individuality, it may be said, in a not very far-fetched sense, that the group of Articulata is composed of compound individuals, whereas the group of Mollusca, hardly excepting the Bryozoa, Tunicata, and Salpæ, consists of single individuals, which possess the utmost uniformity of organization. It is true that the Articulata progress far in that direction; but yet the highest of them, the Insects, do not attain to that singleness of character which exists even in the lower middle ranks of Mollusca. But as this is not the only available character by which we may judge of the point in question, and as there is a more universal and elevated quality, which, under the form of instinct, is so much more highly developed among the immense numbers of the superior ranks, the Insects and Spiders, we can hardly refrain, from a psychical point of view, from classing the Articulata as a whole, if not above, at least fully on a par with the Mollusca.

There is a remarkable correspondence to each other in the respectively repeated parts; for example, the nerve ganglions or knots correspond to the joints of the body; so do the successive chambers of the heart, and the breathing apertures, *stigmata*, or gills, which open along the sides of the joints. This is especially noticeable in the lower kinds of Articulata; but as we ascend toward the more elevated groups, these repetitions are subjected to the same systematic reduction as we have seen operating among Zoöphytes.

Of all the Articulata, perhaps none are so lowly organized, and yet at the same time exemplify the typical idea of this division so fully, as the *Tape-worms*. These I have already given an account of, (p. 83,) and I need, therefore, merely to recur to them here for the sake of bringing their characters into place at the lowest point in the successively rising scale of rank. The much more highly organized *Myrianida*, which I have also described, (p. 80,) will serve as an example of the extent to which the marine worms carry out the idea of serial repetitions, and through a much greater range than Tænia, in fact to the highest degree that is known among Articulata. It might seem to you that on this account Myrianida should rank below Tænia; but the latter is far inferior to the former simply because it almost totally lacks some of the organs which are so highly developed in the marine worm, and therefore the diversity and degree of specialization being also less than in the Myrianida, it of course stands lower in the scale.

*Bonellia.* The most decided step toward that uniformity of organization which culminates in the highest orders of Insects is taken by a group of worms which, curiously enough, until of late years has been classed by almost all naturalists among the Trepangs, — one of the classes of the grand division of Zoöphytes. To this group belongs the worm whose organization I have illustrated by these diagrams (figs. 126, 127). There is scarcely a trace, only a mere rudiment, of the external locomotive append-

216 THE ORGANIZATION

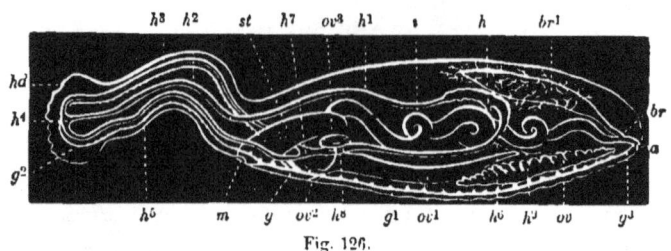

Fig. 126.

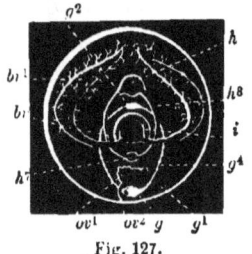

Fig. 127.

ages and gills so numerously repeated along the sides of the body of Myrianida, (fig. 43,) and the interior, with the exception of the string of numerous nervous ganglions, (fig. 126, $g^1$,) presents a uniformity of organization scarcely inferior to that of Insects. When in its native habitat, a cavity in a rock on the sea-shore, it extends its narrower anterior end ($hd$) to an enormous length, as much as eight times as long as the body, and expands the tip sideways, so as to give the whole proboscis the form of a very high **T** with its right and left arms curled more or less backwards.

The mouth ($m$) is not, as one would be apt to suppose, at the

Fig. 126. *Bonellia viridis.* Rol. One half natural size. A diagramic view of a marine worm, in a contracted state. $hd$, the proboscis; $m$, the mouth; $st$, the stomach; $i$, the intestine; $a$, the posterior end of $i$; $br^1$, the respiratory organs; $br$, the point at which $br^1$ communicates with the intestine; $h$, the heart; $h^1$, the median dorsal vessel; $h^2$, the same in the proboscis; $h^4$, point of junction of $h^2$ with the lateral vessels ($h^3$, $h^5$); $h^7$, point of union of $h^5$, $h^3$, under the stomach; $h^8$, the vascular ring about $ov^3$; $h^6$, $h^9$, a small posterior vessel; $g$, $g^1$, $g^3$, the ventral nervous cord; $g^2$, the nervous collar which follows the path of the lateral blood-vessels in the proboscis; $ov$, the reproductive organ; $ov^1$, the matrix; $ov^2$, the outlet of $ov^1$; $ov^3$, the trumpet-shaped entrance to $ov^1$. — *Compiled from the illustrations of Lacaze Duthiers.*

Fig. 127. A diagramic end-view of fig. 126, from behind. $i$, the intestine; $br^1$, the respiratory organs; $br$, point of junction of $br^1$ with $i$; $h$, the heart; $h^3$, lateral blood-vessels; $h^7$, inferior blood-vessel; $g$, the ventral nervous cord; $g^2$, the upper commissure of the lateral branches ($g^1$) of the nervous collar; $ov^1$, the matrix; $ov^2$, outlet of $ov^1$; $g^1$, lateral branches of $g$.

end of the proboscis, but at its base, and exactly in the inferior middle line of the body. Starting from the mouth with considerable breadth (*st*), the digestive canal thins rapidly to a small calibre, and winds in several overarching folds (*i*) alternately from one side to the other, until it comes to a termination (*a*) at the posterior end of the body.

The *respiratory organs*, if such they may be called, are a pair (figs. 126, 127, $br^1$) of spindle-shaped sacs, which are attached on each side of the intestine, and open (at *br*) into it just before its posterior termination (*a*). The whole surface of these sacs is covered by branching tubes, which form so many prolongations from the main cavity upon which they are based.

The *circulatory system*, owing to the enormous extensibility of the proboscis, has a more complicated appearance than really exists. That part of it which most probably corresponds to the heart (*h*) lies in the posterior half of the body, above the intestine, and is to be distinguished from the rest of the system by a considerable thickening at that point, and a puckering of its wall. From the heart the blood is impelled into a vessel ($h^1$) which passes along the middle line of the body and proboscis to the tip of the latter; at this point ($h^4$) it divides right and left, and follows the borders of the T-shaped part along the front and then the back edge, and then courses on each side ($h^3, h^5$) of the main stem to the body. Passing on each side of and below the gullet, close to the mouth, the vessels make a junction ($h^7$) behind the latter, but immediately separate to form a ring ($h^8$) about the trumpet-shaped entrance ($ov^3$) of the emptying conduit, *matrix*, ($ov^1$,) of the reproductive organ, and then unite again into a larger single vessel, which carries the blood along the lower side of the body to the double posterior cavity of the heart. From thence the blood passes upward and forward into the main chamber (*h*) from which it started. If, now, we take a foreshortened view (fig. 127) of this system, we shall see that it is a mere ring about the intestine, with the heart, (*h*,) and its anterior prolongation, above; the two branches along the limbs and stem of the T-shaped proboscis forming the two lateral halves ($h^3$) of the ring; and the main recurrent vessel, ($h^7$,) along the lower middle line of the body, standing at the junction of the vessels which come from the proboscis.

The *nervous system* has the same apparently complicated distribution as the circulatory system, but it is even more simple than the latter. The main portion of it is a thick string ($g$) which rests on the lower middle line of the body, below all the other organs, and extends, with scarcely a diminution in its thickness, from the mouth to the posterior end of the intestine, (from $g$ to $g^3$). At numerous points along its whole length it gives off at right angles, on each side, parallel twigs ($g^1$) which taper into slender threads and bury themselves in the thickness of the highly muscular skin. According to Lacaze Duthiers, from whose figures the diagrams are constructed, there are no ganglionic swellings where these twigs are given off; but yet, inasmuch as we find, in most of the Articulata, such swellings where branches diverge, I think I shall not err if I assume that we have what are essentially the same in Bonellia. Beside these lateral twigs, there are others which branch over the various organs. That part of the system which corresponds to the so-called *brain* and nervous collar of the higher groups, is most singularly disguised in this animal, and is used, it would seem, merely for the purpose of touch. The part in question arises from the anterior end ($g$) of the main ventral cord, and passes, on each side of the throat, into the proboscis. There the two branches ($g^2, g^4$) follow its margin, just within, and below the level of, the line of vessels ($h^3, h^4, h^5$) which run there, and meet along the front of the transverse projections. The latter are said to be extremely sensitive to touch, and apparently in accordance with this, the nervous cord sends off innumerable minute threads which penetrate the skin in the same way that they do in the tactile organs of other animals. Throughout the whole length and breadth of the proboscis the nervous collar preserves a uniform thickness, which is less than one half of that of the main ventral cord ($g$); and there is no part of it which might be called the brain, properly speaking; unless we judge that to be it, from its position, which lies at and about the junction of the lateral halves of the collar, and which exhibits such a high degree of sensitiveness.

The *reproductive system* consists of an egg-bearing portion (*ov*) and the emptying conduit, ($ov^1$, $ov^2$, $ov^3$,) or matrix, as it is called. The former is an elongate narrow mass, with an irregular surface, which stretches along the lower median line of the body, from its posterior end to its middle, and just above the main nervous cord. The *matrix* ($ov^1$) is a hollow, spindle-shaped body which opens ($ov^2$) exteriorly not far behind the mouth, and close to the main nervous cord, ($g$,) either just to the right or the left of it. Not far from this outlet a trumpet-shaped body ($ov^3$) projects from the upper side of the matrix. The trumpet is hollow, and forms a means of communication between the cavity of the body and the interior of the matrix, and thence with the exterior. When, therefore, the eggs are dropped from the ovary (*ov*) they float freely in the body cavity, and in process of time are taken up by the trumpet and passed into the matrix, and by that are cast out, through its inferior opening, ($ov^2$,) into the surrounding element.

The various transitions of form from the lower to the higher groups are so clearly exhibited to the eye among the Articulata, that I would be glad to find time to illustrate all the details of the series of changes which the organs pass through in order to arrive at the most elevated ranks of organization; but I must content myself with a mere indication of some of the great steps, regarding only the more superficial parts of the body.

The *Crustacea*, such as shrimps, lobsters, and crabs, stand next in rank above the Worms. In this group, or *class*, as it is called, we find, as we pass from the simpler to the more highly organized, *i. e.* from the shrimps to the lobsters and thence to the crabs, that the tendency is to mass or concentrate the body in front, and thin it out behind. Those Crustaceans which stand lowest in this class exhibit this tendency by a lengthening of the anterior rings of the body, — as is shown in this figure, (fig. 128,) — forming what is called a head-chest, or *cephalothorax* (*cr*). Eventually, as we ascend the scale, we find this

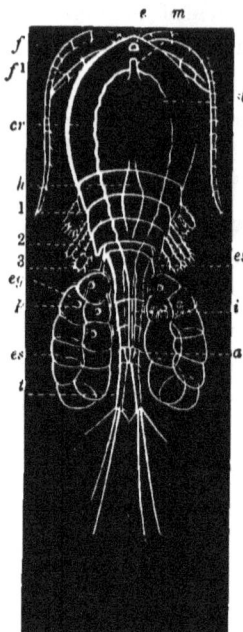

Fig. 128.

cephalothorax becoming a prominent feature and the tail a less conspicuous one; the former gradually extending backwards so as to overlap several of the rings behind it, whilst the latter becomes by degrees bent under the forepart. In this condition you will find the lobster; and in the crab, by the still further carrying out of this *law*,— as the most eminent and profound of American naturalists* has shown it to be,—the tail is almost entirely obscured, and the cephalothorax covers the entire length and breadth of the body.

In the next higher class, the *Arachnida*, or Scorpions and Spiders, the law of centralization and cephalization (tendency toward a head) is carried out in a different form from what appears among Crustacea. Among scorpions the body is divided into two principal regions, a chest, so called, and a tail, but both are elongated and distinctly ringed. In the Spider-group, at least among the highest of them, both the chest (fig. 129, *cr*) and tail, or abdomen (*t*) more properly speaking, are concentrated, and the joints are entirely obliterated. We have thus a distinct specialization of two parts of the animal, a division of the body which foreshadows the more eminent Insects; and in still further confirmation of the tendencies of this group, we find among the highest of them, the Garden Spiders and the like, an initiatory step to sep-

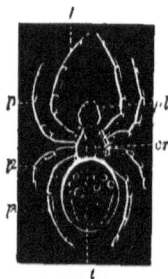

Fig. 129.

\* See J. D. Dana, in *Silliman's American Journal of Science*, 1856, vol. XXII.
Fig. 128. *Cyclops quadricornis*. Müll. 50 diam. A fresh-water, shrimp-like Crustacean, seen from the back. *cr*, the cephalothorax; *t*, the tail; *f*, the

arate that part of the head-chest which embraces the mouth and eyes from the region behind it.

The full realization of this process we find in the free and movable head of *Insects*. They, of all the great division of Articulata, attain to the highest degree of cephalization.

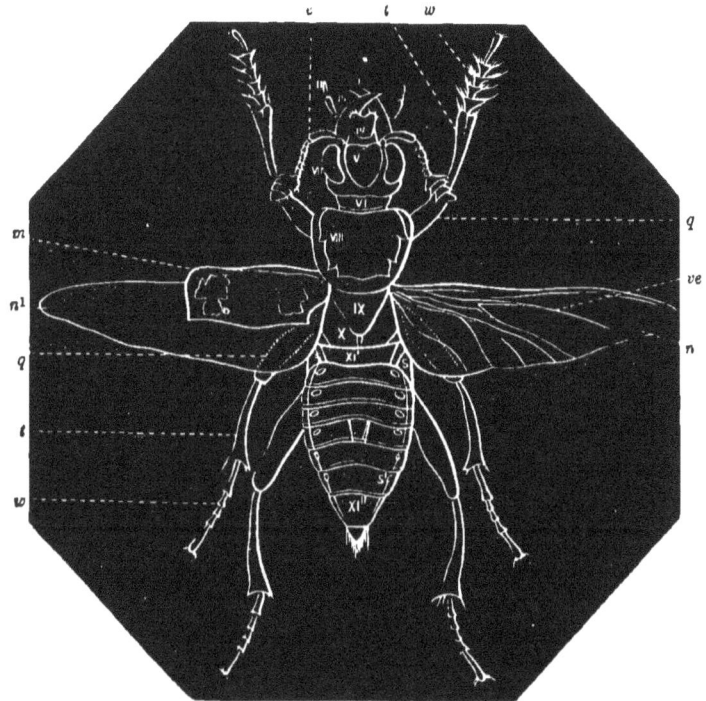

Fig. 130.

This figure (fig. 130) of the common *Carrion-Beetle* will serve larger and $f^1$, the smaller pairs of feelers; 1, 2, 3, the ends of the natatory paddles; *m*, the throat; *st*, the stomach; *i*, the intestine; *a*, posterior end of *i*; *e*, the single median eye; *h*, a heart-like cavity, but not pulsating; *es*, the exterior egg-sacs, partially empty; *es*[1], the funiculus which joins *es* to the body; *eg*, the egg; *p*, the germinal vesicle of *eg*. — *Original*.

Fig. 129. *Epeira trifolium*. Hentz. Natural size. A "garden spider," seen from above. *cr*, the cephalothorax; *t*, the abdomen; *pl*, the feelers, or palpi; *l*, *l*[1], *l*[2], *l*[3], the four pairs of legs. — *From Hentz*.

Fig. 130. *Necrophorus Americanus*. Oliv. Slightly magnified. A common Carrion-Beetle, seen from above. *e*, the antennæ or feelers of the head; I, upper jaws; II, feelers of lower jaws; III, IV, upper lip; V, top of the head;

to illustrate the degree of centralization and cephalization of the lower orders of Insects, and this other one, (fig. 131,) a profile, internal view of a gigantic moth, may typify the perfection of the law. In the carrion-beetle, (fig. 130,) the head (v) is perfectly distinct from the chest, (viii,) but yet it has a broad neck, (vi,) whereas that of the moths, flies, and bees,—insects of the three highest orders, — is a narrow pivot. The chest (*thorax*) of the beetle remains as yet in two quite distinct divisions, (viii and ix, x,) whilst in the moth, (fig. 131,) &c., it is an almost or quite solid piece of concentration. As regards the abdomen (fig. 130, xi' to xi'') of beetles, and the nearly related lower orders, the rings are quite distinct, and among the higher orders they are as a general thing much more consolidated, and in fact so closely united in certain moths, butterflies, bees, hornets, and flies, as to appear to be altogether massed into one.

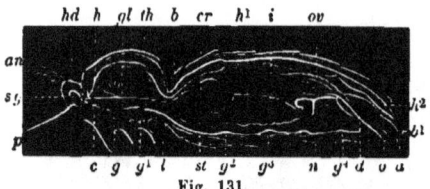
Fig. 131.

The common House-fly is an example of the highest degree to which the law we are speaking of rises, and I am inclined to look upon the order *Diptera*, to which it belongs, as the most eminent

---

vi, the neck; vii, the compound eyes; viii, the prothorax, or first joint of the chest; ix, the scutellum of the mesothorax; x, the third or last joint of the chest; xi' to xi'', the joints of the abdomen; $m$, the left anterior wing; the right one is cut off; $n$, $n^1$, the posterior pair of wings; the vein-like ridges ($ve$) are omitted in $n^1$; $q$, the thigh; $t$, the tibia; $w$, the tarsi, or foot-like part of the leg, terminated by double claws; $s$ to $s^1$, the apertures (spiracles) of the breathing organs. — *From Leconte.*

Fig. 131. *Sphinx Ligustri.* Lin. The Privet Hawk-Moth. Natural size. A longitudinal, sectional view. $an$, antennæ, or feelers; $hd$, the head, or first joint of the body; $th$, the thorax, consisting of the 2d, 3d, and 4th rings; $b$ to $b^1$, the eight rings of the abdomen; $l$, the base of the legs; $p$, the tubular proboscis; $gl$, the gullet; $st$, stomach; $cr$, crop; $i$, intestine; $a$, posterior end of $i$; $h$, $h^1$, $h^2$, heart; $sg$, superior nerve ganglions of head; $g$, $g^1$, ganglions of the thorax; $c$, nervous collar; $n$, main abdominal nerve; $g^2$, $g^3$, $g^4$, ganglions of $n$; $ov$, ovary; $d$, oviduct; $o$, exterior aperture of $d$. — *Slightly altered from Newport.*

of all in the class of Insects. We have, in the peculiar mouth apparatus of the fly, the large, extremely versatile head, the compact thorax, the single pair of wings, and the concentrated abdomen, a series of specializations, reductions to uniformity, and, in fine, the very acme of cephalization, such as are to be found in no other order of Insects.

As an illustration of the internal organization of Insects, I have selected one of the moths, (figs. 131, 132,) known as the hawk-moth, or *Sphinx*. I have already indicated its division into the three distinct regions, head (*hd*), thorax (*th*), and abdomen (*b* to *b*¹). The *head* carries a pair of feelers, *antennæ*, (*an*,) and a tubular proboscis, (*p*,) with which it sucks up the fluid nectar of flowers and various other juices. A pair of compound eyes project from the right and left sides of the head, like those of the carrion-beetle (fig. 130, VII). The *thorax* (*th*) has two pair of wings attached at the sides, and three pair of legs (*l*). The *abdomen* consists of eight inconspicuous rings (*b* to *b*¹).

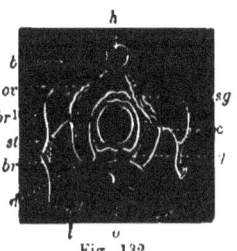

Fig. 132.

The entrance to the *mouth* is between the two slender, furrowed pieces which together form the proboscis (*p*). The *gullet* (*gl*) passes from the mouth in a straight line through the thorax, to the abdomen, and there joins the *stomach* (*st*). The latter is a broad, puckered sac which tapers behind into a considerably convoluted intestine (*i*). Near the junction with the gullet is an oval sac which performs the office of a crop (*cr*). The intestine (*i*) terminates by an opening (*a*) at the extreme posterior end of the abdomen.

The *circulatory system* possesses such an extreme simplicity as to induce some naturalists to place the Spiders, which seem

Fig. 132. An ideal end-view of fig. 131. *b*, the periphery of the body; *l*, the base of the legs; *st*, the intestine; *h*, the heart; *sg*, the super-œsophageal ganglions, or brain; *g*, the sub-œsophageal ganglions; *c*, the nervous collar; *or*, the pair of reproductive organs; *d*, the oviducts; *o*, aperture of *d*; *br*¹, the respiratory tubes, *tracheæ*; *br*, external aperture, *spiracle*, of *br*¹. — *Original.*

to have a more complicated system, above Insects. But the truth is, the blood circulation in Spiders is not so intricate nor so highly specialized, when compared with that of Insects, as has been asserted. In Insects the only definitely circumscribed canal of the system is the so-called *dorsal vessel*, or *heart* ($h$, $h^1$, $h^2$). It is a mere tube which lies close to the back, in the middle line, and extends from the head to the posterior end of the abdomen. It is open at both ends, ($h$, $h^2$,) and at the successive points where it is narrowed are a pair of apertures, one on each side, which are guarded by a valve on the inside. There is also a pair of internal valves at the posterior opening. The process of circulation is a very simple one: the blood enters the posterior and lateral apertures of the heart as it expands; then upon its contraction all the valves are closed and the blood is forced toward the head, and, passing out at the anterior opening, flows, in numerous currents, and, at first appearances, in undetermined channels, among the various organs, and into the legs and wings. A careful examination of some of the more transparent insects, such as the May-fly, (*Ephemera*,) Gall-fly, (*Cynips*,) Plant-louse, (*Aphis*,) Lace-winged fly, (*Chrysopa*,) Dragon-fly, (*Æshna*, *Agrion*, *Libellula*,) and the grub or worm of many more, has convinced me, that, notwithstanding the apparent lack of walls to the channels of circulation, the course of the blood is none the less definite; always passing in one set of channels going from the heart, and returning toward it in another set. This is particularly noticeable in the head, legs, and wings.

The *breathing organs* consist of numerous branching tubes, *tracheæ*, (fig. 132, $br^1$,) which spread out from certain fixed points along the right and left sides of the body. The air enters the body through minute apertures, *spiracles*, ($br$,) of which there are two in each ring, one on each side, as in this beetle (fig. 130, $s$ to $s^1$). At each spiracle (fig. 132, $br$) a tube ($br^1$) arises and branches into innumerable twigs, which spread themselves over the various internal organs. The walls of the tracheæ being very thin, the contained air finds a ready absorbent in the circulating fluid.

The *nervous system* consists of a series of swellings, *ganglions*, united by threads, which extend from the head to the posterior end of the body, along its lower median line. In the head is a double ganglion, (*sg*,) the so-called *brain*, which lies across the upper side of the throat (*gl*) just behind the mouth. From the " brain " a thread passes around on each side of the throat to the lower side, and unites with the first ganglion (*g*) of the thorax. This constitutes the nervous collar, (*c*,) and serves to unite that part of the system above the intestinal tract to that below it. The anterior (fig. 131, *g*) of the thorax-ganglions is scarcely separate from the succeeding one, and the latter ($g^1$) is the result of the intimate fusion of the second and third. Their consolidation corresponds to the degree of concentration that the rings of the thorax attain to; there being in the latter an imperfect separation between the first and second joints. The abdominal ganglionic chain (*n*) is single, and passes backwards in a direct line from the posterior double thoracic ganglion, with a ganglion ($g^2, g^3, g^4$) at nearly every joint. From these ganglions minute nerves pass off, right and left, above them and on each side to the various organs; and likewise similar nerves, arising from the " brain," supply the feelers, expand in the eyes, and branch over the upper side of the throat and stomach.

The *reproductive system* lies in the posterior part of the abdomen. It consists of a pair of bunches (*ov*) of wavy tubes, which converge into one channel, *oviduct*, (*d*,) on each side. The two oviducts pass down each side of the intestine to a point just in front of its termination, where they unite in one common outlet (*o*) on the middle, inferior line of the body. The eggs are generated in the bunches of tubules, *ovaries*, (*ov*) ; and when ripe they pass into the oviducts, and thence through the common channel (*o*) to the place of deposit.

## CHAPTER XIII.

#### VERTEBRATA.

As I have already and at considerable length (p. 124) discussed the characteristic features of *Vertebrata*, I need not enter into any further details in regard to their relation to the *typical* form, but simply ask you to make the comparison for yourselves between the ideal figures (figs. 63, 64) of this type and the diagrams of the actual organization which I am about to describe. I hardly need to say that you will not find an identity in the details, but a perfect accordance in the relative position of the four systems of organs, namely, the nervous, vertebral, digestive, and circulatory. Among the lower groups of Vertebrata these systems are more clearly demonstrable in a diagram than among the higher ranks, as you may see by a comparison of this diagramic illustration of a Fish (fig. 133)

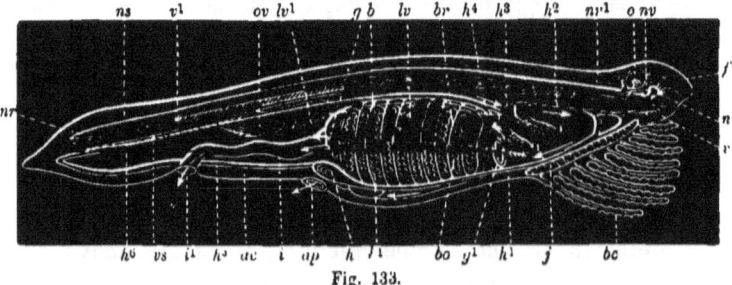

Fig. 133.

Fig. 133. *Amphioxus lanceolatus*. A diagramic figure of the Lancelet. Natural size. $f$, the head; $r$, $v^1$, the notochord, or vertebral column; $vs$, the sheath of $r$, $v^1$; $bc$, the buccal cirrhi; $j$, the buccal ring at the entrance to the mouth; I, II, oval bodies projecting freely into the buccal cavity; $bo$, the lateral branchial openings; $g^1$, entrance to the throat or branchial cavity; $g$, posterior end of the latter, and entrance to the intestine proper ($i$): $i^1$, posterior end of $i$; $lv$, $lv^1$, appendage to $i$, opening into it at $lv^1$; $h$, the heart; $h^1$, $h^2$, the anterior blood-vessels; $h^3$, recurrent branches of those, from $h^4$, which supply I, II; $h^5$, $h^6$, the

with the *ideal* Vertebrate, (fig. 63,) when both are contrasted with the more complicated, warm-blooded quadruped (fig. 134).

*Amphioxus.* The most lowly organized of all known Vertebrates is a fish which is commonly called the Lancelet, and sometimes the Sand-Eel, on account of its habit of burrowing in the sand of the sea-shore. It is so remarkably transparent that its whole internal organization can be seen with a good microscope. It has no external appendages excepting a circle of feelers (fig. 133, *bc*) about the mouth, (*j*,) and therefore, for the lack of fins, the whole duty of locomotion devolves upon the highly muscular, lance-like tail. The *head* (*f*) is peculiarly adapted, by its sharp, thin front, for the purpose of penetrating the compact sand-beach. The mouth (*j*) is an elongate opening situated on the under side of the head, at a considerable distance behind its front. The feelers (*bc*) which surround it are largely supplied with nerves, and therefore in all probability are highly sensitive organs of touch, and serve as efficient means for obtaining food.

Between the mouth and the entrance ($g^1$) to the throat, there is considerable space, the *buccal cavity*, within which certain oval bodies (I, II) project from above like a pair of palates. The latter are covered by constantly vibrating threads, *cilia*, which keep up a current of water from the mouth toward the gills, (*b* to $b^1$,) and at the same time furnish the means of floating fine particles of food into the throat. The entrance to the latter is a moderate aperture, ($g^1$,) but the throat itself is a very large cavity, which performs at the same time the office of a breathing apparatus, or *gill-chamber.* Its sides are perforated by several parallel slits, (*bo*,) which extend from its upper (*b*) to its lower ($b^1$) margins, through which the water pours, as between the

---

dorsal artery; $h^5$, the abdominal vessel; *b*, the upper, and $b^1$, the lower point of junction of the branchial vessels (*br*) with the dorsal ($h^1$, $h^6$) and ventral ($h^1$, $h^5$) vessels; *ac*, abdominal cavity; *ap*, abdominal pore; $nr^1$, the anterior, and *nr*, the posterior end of the main nerve or spinal marrow; *ns*, sheath of *nr*, $nr^1$; *o*, the eye; *n*, the olfactory nerve; *nv*, the facial nerves; *ov*, the reproductive organ. — *From Owen*.

gills of ordinary fishes, into the general cavity (*ac*) of the body, and thence passes out of the *abdominal pore*, (*ap*,) to the exterior world. The true intestine (*i*) commences with a moderate aperture (*g*) at the posterior end of the branchial gullet, and proceeds to its termination ($i^1$) in a nearly direct line. Near its beginning a saccular organ (*lv*, $lv^1$) which is thought to perform the office of a liver, opens into it.

The *circulatory system* can hardly be said to possess a heart. The oval enlargement (*h*) in the diagram is an exaggeration, merely to render conspicuous the position in which the heart belongs. The whole system of vessels is said, by those who have examined the animal in a living state, to contract from one end to the other. According to this diagram the blood courses from the central cavity, *heart*, (*h*,) in a backward direction, in a single vessel ($h^5$) along the lower side of the body to the tip of the tail, and then, doubling upon itself, it passes along the upper median line ($h^6$ to $h^4$) of the general cavity of the body, and directly over the gill-chamber (*b* to $b^1$) into the head. At the fore part of the gill-chamber it forks (at $h^4$) and sends a branch into each of the ciliated palate-like bodies (I, II) of the buccal cavity; and after penetrating to the tips of these, the two branches double upon themselves, and, returning, ($h^3$,) unite again into one vessel ($h^2$). This continues its course along the dorsal median line ($h^2$) further into the head to a point over the anterior edge of the mouth, and there, by two vessels, joins the return current, ($h^1$,) which passes along the lower middle line of the gill-chamber to the heart. The circulation is further complicated by intermediate currents which pass from the upper median vessel ($h^4$) directly downwards in smaller channels (*br*) which lie between the slits (*bo*) of the gills, and empty into the lower median vessel ($h^1$) where it joins the stream coming from the head. In full-grown specimens of the Lancelet there are as many as fifty of these transverse *branchial vessels*.

The more recent investigations of Quatrefages differ* in their

* Quatrefages, Voyage en Sicile, vol. II. p. 12, and Annales des Sciences Naturelles, 1845, vol. IV.

results from what is given by Owen, inasmuch as the former represents the currents as passing from the heart ($h$) partly *forward* into the head, and in part through the branchial vessels ($b$, $b^1$, $br$) *upward* to join the dorsal vessel ($h^4$, $h^6$) bearing the current from the head toward the tail, and from the latter returning to the heart in the lower median vessel, ($h^5$,) which passes along the inferior side of the intestine.

The *vertebral column* or spine is represented by a mere cord ($v$, $v^1$) of jelly-like matter, which extends along the middle of the back, from the front ($f$) to the posterior end of the tail. It is enclosed in a membranous sheath, ($vs$,) within which it lies so loosely that it can be taken out entire by simply opening a gash along the back. It usually goes by the name of the *notochord*, and corresponds with the first rudiment of the spinal column of the embryo of higher animals.*

The *nervous system* consists of a main cord ($nr$, $nr^1$) and numerous branching prolongations which project right and left from its sides. At its anterior end ($nr^1$) there is no sensible dilatation which corresponds to the brain. The nerves of the eye ($o$) and nose ($n$) arise from near its rounded end in the same simple way as the other less specialized nerves ($nv$) which spring up near them and branch in the head.

The *reproductive system* is a mere elongated oval mass ($ov$) attached to the upper median line of the general cavity, just behind the branchial chamber. The eggs reach the outer world by dropping from the *ovary* into the visceral cavity and thence passing out through the *abdominal pore*, ($ap$,) an aperture which lies in the lower side just behind the heart.

If now I have given you a clear understanding of the relation and nature of the organs of the Lancelet, there will be no difficulty in comprehending, by the help of a few words of explanation, the organization of one of the highest of the Vertebrates. With the preliminary knowledge of the simpler structure of the one, the apparently puzzling complication of the other may be

* See the embryology of the Turtle, in chap. XVII., where this body is called the *chorda dorsalis*.

resolved at a glance. In the Lancelet, and in all fishes, and even in the lower grades of reptiles or reptilian fishes, (Lepidosiren, chap. XVI.,) the head is not distinctly separated from the body; but, as we ascend the scale, through the groups of the higher reptiles, (Lizards, Turtles, &c.,) birds, and finally the warm-blooded vertebrates, Mammals, (figs. 26, 134,) the process of cephalization becomes more and more clearly marked by an external configuration, and within the head by a concentration of the regions of the brain, and the growing preponderance of the divisions devoted to the more delicate sensations of sight and reasoning.

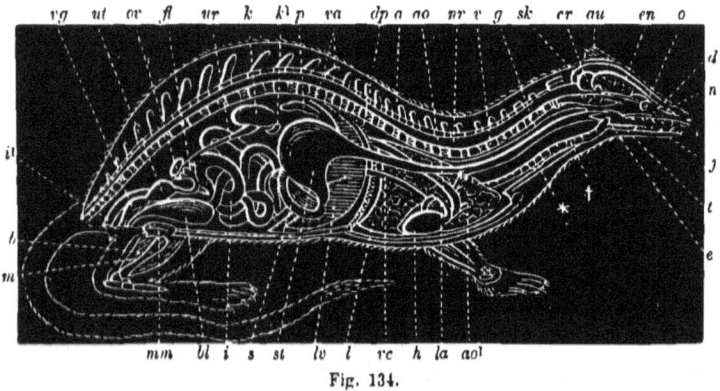

Fig. 134.

*Mammalia.* — The body of the higher Vertebrates is divided into head, chest, and abdomen; the first contains the brain (fig.

Fig. 134. A diagramic longitudinal section of a Mammal. *sk*, skull; *v*, vertebræ; *a*, dorsal arches of the vertebræ; *va*, the upper and lower portions of the vertebral arch; *j*, lower jaw; *b*, bone of the leg; *m*, muscle; *d*, teeth; *t*, tongue; *g*, gullet; *, thyroid gland; *st*, stomach; *i*, intestine; *i¹*, end of *i*; *lv*, liver; *p*, pancreas; *s*, spleen; *k*, kidneys; *k¹*, appendages to *k*, known as the suprarenal capsules; *ur*, outlet of *k*; *bl*, bladder; *e*, epiglottis, or entrance to the windpipe (†); *l*, lung; *h*, heart; *ao*, abdominal aorta; *ao¹*, carotid artery going to the head; *vc*, vena cava inferior, or abdominal vein; *la*, pulmonic artery; *dp*, diaphragm; *o*, the eye; *en*, cerebrum; *cr*, cerebellum; *n*, olfactory nerve; *au*, the outer ear; *nr*, spinal marrow, or main nervous cord; *ov*, the ovary, or eggbearing portion of the reproductive organ; *fl*, the trumpet-shaped Fallopian tube through which the eggs pass into the uterus (*ut*); *vg*, the vagina, or outlet of *ut*; *mm*, the mammæ, or milk-bag. — *From Owen. Slightly altered.*

134, *en, cr*) and the organs of sensation (*o, n, t*); the second encloses the heart (*h*) and lungs (*l*), and the third, which is separated from the second by a transverse membranous, muscular partition, the diaphragm, (*dp,*) is occupied by the stomach, (*st,*) intestines, (*i,*) liver, (*lv,*) kidneys, (*k,*) spleen, (*s,*) bladder, (*bl,*) and the organs of reproduction, (*or, fl, ut, vg*). The first, the head, is in great part an expansion of the spinal column, (*v, va,*) under the guise of the skull, (*sk,*) and has appended to it the jaws, (*j,*) between which and the skull is the mouth, the enclosed tongue, (*t,*) and the entrance to the throat (*g*) and windpipe (*e,* †).

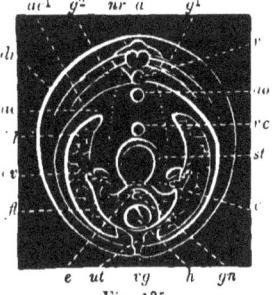

Fig. 135.

The *spinal column* (*v, va, a*) consists of a longitudinal series of bones which, as it were, overarch the organs of the chest and abdomen and continue into the tail. Each bone of the series consist of a central portion, the *centrum*, (*v,*) and a hollow arch (*a*) above the latter. Within this arch the great nervous cord (*en, cr, nr*) runs from the skull, which is the first arch, to the tip of the tail. The ribs and bones of the limbs (*b*) are lateral appendages of the vertebral column, to which they are attached by ligaments, and upon which they are moved backward and forward by muscles (*m*) which surround them. We have by this arrangement what appears to be a separation of the body into two distinct superposed cavities, the one containing the centre of the nervous system, and the other the visceral organs; but if now we turn back to the Lancelet, (fig. 133,) I do not think we shall revive any impression of such a state of things existing there. In the latter case we have a mere gelatinous cord (fig. 133, *v, v¹*) underlying the nervous cord (*nr, nr¹*); the

Fig. 135. An ideal foreshortened view of 134, as if seen from behind. In addition to the letters in fig. 134, there are *dr*, the skin; *ac¹*, the periphery of the abdominal cavity (*ac*); $g^1$, the posterior pair of sensory nerves; $g^2$, the anterior pair of motory nerves; *c*, the *glosso-pharyngeal* nerve, forming in part the nerve of taste, which is concentrated (at *gn*) under the tongue. — *Original*.

former evidently having no sort of relation to the latter, as regards its connection with the other organs, in a functional sense, but purely one of position. Not only is this true of the Lancelet, but also of many other nearly related fishes, such as Lamprey-Eels, Myxine, and certain Sharks; and it is not until, by following up the series, we arrive among the considerably more elevated forms, that we find the arch completely encircling the nervous cord; and not even in the whole class of fishes are the successive arches so close together as to lead any one but an over-eager advocate of such a view to conceive that the series of perforations which they enclose are functionally a closed cavity. The open spaces between the successive arches are much wider than the spaces which they enclose, and it is only in the highest animals that this proportion is reversed. Yet even in the latter case, the interspaces are by no means closed ones, but are open to a greater or lesser extent to allow the passage of the great lateral nerves (fig. 135, $g^1$, $g^2$) which diverge right and left from the main cord ($nr$) and branch through the various organs.*

The *digestive system* comprises the mouth, teeth, ($d$,) a slender gullet ($g$) leading through the diaphragm ($dp$) to an oval expansion, the paunch or stomach, ($st$,) then contracting in a long, convoluted intestine ($i$) which terminates ($i^1$) at the posterior end of the body. Appended to the fore part of the intestine, just behind the stomach, is the large concavo-convex liver, ($lv$,) and close to it the spleen ($s$). Near the back are the kidneys, ($k$,) which filter off the waste fluid of the body, and through slender tubes ($ur$) pour it into the bladder ($bl$).

* Should any one urge that the vertebral arches embrace what is to all intents and purposes a closed cavity, as contradistinguished from the visceral chamber, and that in the lower Vertebrates the mere rudiments of the arches are sufficient indications of the universal presence of such a cavity, I am willing to admit the truth of the assertion, *provided* that I may claim at the same time that the cartilaginous box of the head of Cephalopoda, and the more or less complete arches which border the main nervous cord of Insects, form what is essentially a similar segregated cavity. The basis of difference between the Vertebrata and the other grand divisions lies in the *notochord*, and not in the presence of an imaginary upper and lower cavity.

The *circulatory system* has its starting-point in a highly muscular, fourfold sac, the *heart*, ($h$,) which lies about in the median line of the chest, below the gullet ($g$) and close to the breastbone. The principal vessels connected with it are the arteries which carry the blood to the head, lungs, and posterior regions of the body, and the veins in which the return currents bring back that fluid to the heart. The course of the blood in completing its circuit is rather complicated, if we follow it in all its details, but the essential lines of travel are quite simple in their connections. Starting from one of the chambers of the heart, which is called the *left ventricle*, the blood passes, through the *great artery*, ($ao$,) a short distance forward, and then, as it turns upward, a smaller current diverges toward the head in what is called the *carotid artery*, ($ao^1$,) whilst the main current ($ao$) continues up to the lower face of the spine and follows its line backwards in what is known at this part of its course as the *abdominal artery*. From the latter numerous vessels are given off to the various visceral organs, and from the *carotid* ($ao^1$) the organs in the neck and head are supplied with branches. The *return currents* start in the minute, branching, capillary vessels which are continuous with the equally minute, branching terminations of the arteries. The capillary return vessels, *veins*, gradually unite in fewer and larger vessels, which finally coalesce in one large vein in front, coming from the head, and in another great vein behind, ($vc$,) coming from the abdomen. Each vein, the *vena cava superior*, and the *vena cava inferior* ($vc$), empties its contents separately into the *right auricle* of the heart, and thus one branch of the circuit is completed. The other branch of the circulation is devoted purely to the aëration of the blood. The venous blood passes from the chamber in which it was received, *i. e.* the right auricle, into the *right ventricle*, and from thence it is thrown through the *pulmonic artery* ($la$) into the minute branching vessels of the lungs ($l$). After being aërated there, the blood flows through the capillary veinlets into the great recurrent vessels, the *pulmonary veins*, from each lung, and thence enters the *left auricle*, and finally completes its tour by passing into the *left ventricle*, the same from which it went out in its first circuit.

The *organs of respiration* are a windpipe and a pair of lungs. The *windpipe* (†) opens (at *c*) at the root of the tongue, and after passing backward a considerable distance, and in front of the gullet, (*g*,) it forks into two branches, of which one goes to the right and the other to the left lung. The *lungs* (*l*) are great sacs divided into numerous, irregular compartments, like the interstices of a sponge, in the thickness of whose meshes the minute arteries and veins pass and repass, to carry the blood from and to the heart, during the process of aërification.

The *nervous system* I have already stated to lie in a partially overarched furrow upon the back of the spinal column. The anterior part of it, the *main organ*, (*en, cr,*) is enclosed within the cranium, (*sk,*) and consists of two great double masses, whose halves lie symmetrically right and left of the median line of the head. The anterior mass, the *cerebrum*, (*en,*) is by far the greater of the two, and from it proceed the nerves of the two most delicate senses, namely, sight (*o*) and smell (*n*). The posterior mass, the *cerebellum*, (*cr,*) is completely overlapped by the cerebrum in man, but projects beyond it in the monkeys and the groups below them. The nerves of hearing and of taste (fig. 135, *gn, c*) arise from the *medulla oblongata*, that part of the brain which underlies the cerebellum (*cr*) and forms the immediate transition to the main cord, (*nr,*) or spinal marrow, which joins it at the base of the skull. The nerves which control the motions of the body, the *motory nerves*, (fig. 135, $g^2$,) and the nerves of *sensation*, ($g^1$,) originate in pairs from the right and left sides of the spinal marrow, (*nr,*) along its whole length.

The *reproductive organs* lie symmetrically right and left of the axis of the body, and consist of a pair of small, oval, egg-bearing organs, the *ovaries*, (*ov,*) and two outgoing conduits (*fl*) which unite in one common receptacle, the *uterus* (*ut*). When the eggs are ripe, they drop from the ovary (*ov*) into the trumpet-shaped mouth (*fl*) of the conduit, *Fallopian tube*, and thence are conducted to the uterus (*ut*). There they undergo a change, and, taking on the form of the parent, remain until they have attained to a certain degree of development, and then make their exit through the vaginal outlet (*vg*). For the nourishment of their young the warm-blooded Vertebrates possess an organ for the secretion of milk, which is called the *mammary gland*, (*mm*,) and which, in such animals as the cow, sheep, and goat, is known as the *milk-bag*, and either hangs, as in most of the quadrupeds, between the hind legs, or extends to a greater or less distance along the belly. Among the higher groups the mammæ lie between the fore limbs.

## CHAPTER XIV.

#### THE DISTINCTNESS OF THE FIVE GRAND DIVISIONS.

I THINK I have said enough already to convince you that I do not advocate the possibility of a *transition* from one grand division to another; and it is here that I shall probably disappoint many who believe that there are no distinct types of form in the animal kingdom. Even Lamarck advocated the distinctness of certain types, and moreover he expressed a belief in the possibility of discovering others. That there are certain *ideal transitions* or *relations* between the *five grand divisions*, you must have learned from the description of these diagrams of the ideal types (figs. 55 to 64). Life is the first ideal relation among them that impresses us; then there is *bilaterality*, then *bipolarity*, *i. e.*, the opposition of *above* and *below*, and in more or less intimate relation with this is the *tendency* of the *nervous system* to condense toward the side opposite the heart, and at the same time to preponderate in the regions next the head. All these are *ideal*, and are the expressions of a *gradational* progress from a lower to a higher type; from an idea of low, undetermined, diffuse organization, as in Protozoa, to the highest degree of *specialization* and concentration, such as is to be seen in the Vertebrates as a whole, and in *Man;* in whom the crowning effort was centered on the brain, the throne of thought, before which all the other faculties stand in inferior ranks.

Now if these which I have called "*ideal relations* between the great types" were not so, but rather real relations, *i. e.*, *relations of consanguinity*, we ought to be able to trace the organization of an animal of *one type* into that of *another;* and in order now to show that this cannot be done, I shall proceed to make comparisons between the types of the grand divisions, as if I were trying to prove their identity; and in this way I shall be able to make the closest comparison possible. For the sake of conformity to my previous method of procedure, I begin with the lowest types, and from them rise to the higher forms.

*Protozoa and Zoöphyta.* The first on my list is the alleged transition from the *Protozoa* to *Zoöphyta.* Among Protozoa the *Vorticellæ* (Epistylis, p. 161, fig. 95) are those which have the closest resemblance to Zoöphyta; and among Zoöphyta, *Hydra* (p. 55, fig. 27) most resembles Vorticellæ. Let us place these two forms side by side and try to find what are their points of relationship, if they have any. This can be most easily accomplished by recurring to the descriptions which I have already given of their internal structure. The *Epistylis*, or " Bell-animalcule," as I have shown, (p. 161, fig. 95,) possesses an oblique mouth, ($m$,) a spiral gullet, ($g$, $g^1$,) a one-sided digestive cavity, ($m$ to $s$,) a peculiar pulsating sac, or contractile vesicle, ($cv$,) and an internal organ of reproduction ($n$). On the other hand, I have described (p. 55, fig. 27) *Hydra* as a simple sac or tube whose mouth opens directly into a wide digestive cavity, ($s$,) which occupies the whole length and breadth of the body, to the exclusion of any other organ whatever.

The *Epistylis* has a type of organization which unmistakably allies it with such animals as *Zoöthamnium*, (p. 175, fig. 104,) *Stentor*, (p. 62, fig. 30,) *Paramecium*, (p. 163, fig. 96,) *Pleuronema*, (p. 170, fig. 99,) *Dysteria*, (p. 171, fig. 100,) &c., all of which are connected in a common type by their spiral contour, oblique mouth, one-sided digestive cavity, contractile vesicle, and a peculiar, internal, reproductive organ. There are certain of the Protozoa which would appear to be an exception to the typical oblique form; such an one is represented in this figure (*Podophrya*, p. 51, fig. 25). It is an irregular, four-sided, inverted pyramid, whose apex rests upon a stalk, and whose four corners are occupied by a group of feelers ($f$); and within are the characteristic contractile vesicle ($cv$) and reproductive organ ($n$). This is its conformation in its adult stage, when, as is frequently the case with many other animals, it assumes a disguised shape, and belies the relationship which its internal organization reveals. In its younger stages (fig. 25, A, B) its exterior form possesses the characteristic obliquity of the type to which it unquestionably belongs.

In the *Hydra*, on the contrary, we find altogether different tendencies. What are the peculiarities of the type to which it belongs I have already (p. 177) described, when speaking of the organization of Zoöphytes; but I wish at this point to say a few words in explanation of those nearly related Hydroids which form the connecting link between it and the more complicated forms. The first one that I shall take up is interesting because it not only illustrates this transition, but also displays a bilateral character in a most unquestionable light. It is a Hydroid by the name of *Tubularia*, (fig. 136,) and is common on the whole North Atlantic coast, both in Europe and America. The cylindrical stem ($s$ to $s^1$) supports a distinct head, ($hd$,) which at the base bears a wreath of twenty to twenty-five tentacles, ($t$,) and projects into a broad, cylindrical proboscis, ($p$,) whose end is pierced by the mouth and covered by several rows of short tentacles ($t^1$, $t^2$). Upon cutting across the stem, we find that it does not embrace a single cavity like that of the Hydra, but that its centre (fig. 137, $c$) is a

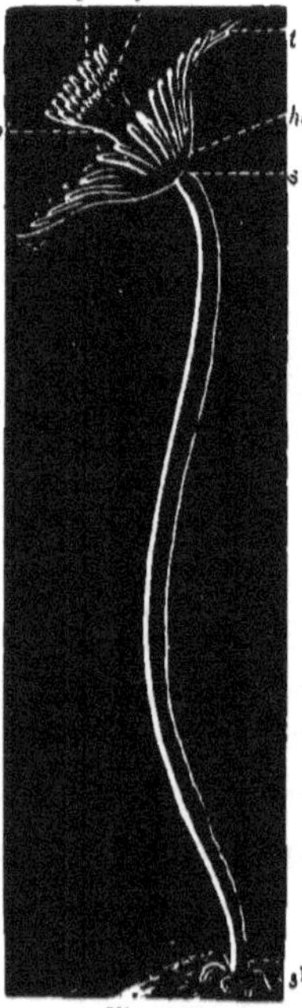

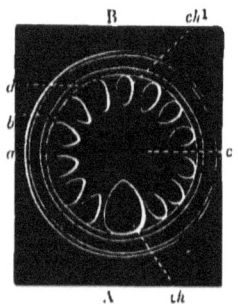
Fig. 137.

Fig. 136.

Fig. 136. *Tubularia indivisa*. Lin. Natural size. Marine. From Boston Harbor. $s$ to $s^1$, the stem; $hd$, the head; $t$, the posterior tentacles, or *corona*; $t^1$, $t^2$, the tentacles of the proboscis ($p$). — Original.

Fig. 137. An actual transverse section of the upper part of the stem of fig.

solid mass of cells, and its periphery is occupied by several longitudinally arranged tubes ($ch$, $ch^1$). One ($ch$) of these tubes is much larger than the others; and if we trace it toward the posterior end of the stem, it appears as the original one from which the others have arisen right and left. It is, therefore, the dominant feature of the stem, and constitutes the basis upon which to estimate the relations of the regions about it. A line drawn through it to the opposite side (through A, B) divides the stem into symmetrical right and left parts; that is to say, the stem is *bilateral*, and its periphery is occupied by several laterally repeated tubes.

The crown of tentacles (fig. 136, $t$) at the base of the head is not, as it appears to be, a single row, but is composed of at least three circles placed one before the other relatively; but yet as the feelers of each circle alternate with those in the others, all three circles are enabled to overlap each other so closely as to produce the appearance of a single one. If, however, their development is watched during the earlier stages of growth, their separate origin can be easily verified.*

The manner in which the tentacles of the proboscis comport themselves now and then illustrates, in a very apt way, their

136. 12 diam. $a$, the parchmenty sheath; $b$, the outer wall; $d$, the inner wall; $c$, the solid core; $ch$, $ch^1$, the channels at the periphery of $c$. — *Original.*

* This diagram (fig. 138) will serve to illustrate the manner in which the tentacles of the crown originate and change their position during the process of development of the young. Tubularia is usually born with eight tentacles in the corona. These may be represented by those in the centre (I, I, A, D) of the figure. Next, eight more (II, II, B, E) begin to develop farther back than those of the first group, but as they grow they gradually press forward and fill the intervals between the latter (*i. e.* as if $b$ were to follow the direction indicated by the arrow). Meanwhile a third group (III, III, F, C) of sixteen springs up still farther back than the second, each tentacle (III) alternating with those of the first (I) and second (II) groups, and in process of time they too push forward and occupy the intervals between the members of the first and second groups (*i. e.* as if $a$ and $c$ were to follow the directions indicated by the arrows). This is the way in which the tentacles (fig. 136, $t$) of the corona of an adult head happen to appear to be in a single circle. Upon close examination, however, even of a very old head, one may detect a difference in the relative level of these tentacles.

arrangement in the crown. Usually the proboscidal tentacles stand in three or four transverse circles, the older ones ($t^1$) nearest the end, and the youngest ($t^2$) the farthest from it; but occasionally the mouth of the proboscis expands very widely, like the end of a trumpet, and at the same time decreases its length, by which the alternating tentacles of the three circles are reduced to one level, so as to lie in a single unbroken circle. The change of place which they undergo in order to arrive at this position is precisely that to which the feelers of the *corona*, at the base of

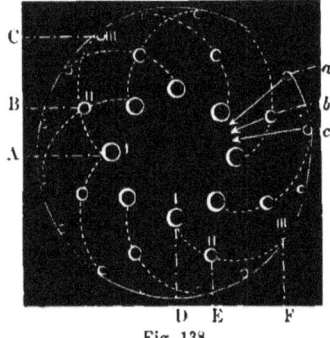

Fig. 138.

the head, are subjected as they successively develop in their respective circles. This diagram (fig. 138) presents an end-view of three circles of tentacles, and is intended to illustrate their relative position at the end of the proboscis. The inner circle represents the larger ones (I) nearest the mouth; those of the next outer circle (II) alternate with the first; and in the third or outermost circle they (III) alternate with those of both inner circles. They appear, therefore, to be arranged in spirals, which, as the dotted lines indicate, may wind either to the right or to the left; but a study of their process of development discloses the fact that their arrangement is simply one of alternation, and not that of a true spiral. In order to bring the three alternating groups into one, the tentacles of the two outer circles (II, III) have but to move in direct lines toward the centre, as the arrows indicate; or, what is more frequently the case, reversing the direction of the arrows, the two inner (I, II) must move outwardly.

Fig. 138. Diagramic representation of the relative position of the tentacles of the proboscis of Tubularia, (fig. 136, *p*,) and serving to illustrate the mode of development of the tentacles of the corona (fig. 136, *t*). A, B, C, D, E, F, the same as I, II, III, for which see the body of the work. — *Original*.

Fig. 139. The Scyphostoma of *Aurelia flavidula*. Per and Les. Magnified 8 diameters. The ends of the tentacles (*t*) are left out of view (see p. 67, fig.

The tentacles (fig. 139, $t$) of the *Scyphostoma* of *Aurelia* (p. 70, fig. 37) seem to be in a single circle, whereas they have the same arrangement as those of the corona of *Tubularia*. The internal organization of the Scyphostoma, as well as the mode of development of its tentacles, has an unquestionable relation to right and left. At a very early period the interior is subdivided into four longitudinal, partially separated compartments, by the projection of four equidistant, thick semi-partitions. In full-grown individuals (fig. 139) these semi-partitions (fig. 140, $a$,) extend from the base of the tentacles ($t$, $t^1$) halfway to the bottom of the general cavity ($gc$). On account of their transparency, they have been mistaken for tubes, and the fold ($b$, $b^1$) opposite each one of them, in the muscular layer, has also been erroneously described* as a channel in which the fluids circulate; but each of the first ($a$) is a solid, jelly-like mass standing out like a *pilaster*, and the second ($b$) is so doubled upon itself longitudinally as to present a quadruple contour, like the two outer and two inner profiles of a tube. The thickness of the walls at the base ($d$) of the circle of tentacles has also the appearance of a circular tube into which the four supposed lon-

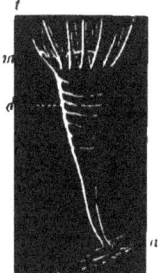

Fig. 139.

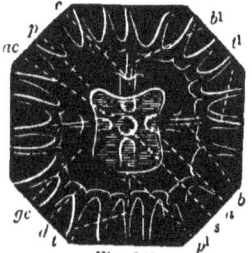

Fig. 140.

34). $a$, the base; $m$, the partially extended proboscis, encompassed by the tentacles ($t$); $d$, the incipient formation of the discs of the Ephyra, foreshadowed by transverse rings. — *Original*.

Fig. 140. An end-view of the head of fig. 139. 20 diam. $t$, $t^1$, the tentacles; $p$, $p^1$, the margin of the four-sided proboscis; $gc$, the cavity of the body; $ac$, the bottom of $gc$, where the cavity ($s$) of the stem opens; $d$, the periphery of $gc$, at the base of the tentacles; $a$, the semi-partitions, or *pilasters*; $c$, the depression immediately over the anterior end of each pilaster, which appears like an aperture; $b$, $b^1$, the longitudinal fold. — *Original*.

* See Frantzius on the Scyphostoma of *Cephea borbonica*, in Zeitsch. für wissenschaft. Zoölogie, June, 1852. Also Gegenbaur upon the same, in his Generationswechsel, 1854, Taf. II. fig. 34.

gitudinal tubes seem to empty. Such is one of the curious fallacies into which naturalists have fallen in their desire to homologize in a special manner the organization of the Scyphostoma with that of the full-grown Aurelia.

Now when we trace the development of the tentacles, we find that they arise at points which have definite relations with the position of the four equidistant pilasters (*a*). In the first stage of growth, after the egg-phase, the embryo has a cylindrical  form, like this, (fig. 140*a*,) and at one end a simple aperture (*m*) for a mouth, which leads to an elongate cavity (*d*) within. After swimming about for a while by means of its vibratile cilia, it settles down on the end opposite to the mouth, and becomes fastened to whatever it rests upon. Soon after this, the four pilasters, which I just spoke of, make their appearance within the elongate cavity of the cylinder, and about the same time, the tentacles begin to develop. At first, either two, as in this figure, (fig. 140*b*, *t*,) or at most four tentacles, spring up at as many alternate points with the pilasters; then four others, either consentaneously or alternately, originate opposite the pilasters; and the succeeding ones, in successively overlapping circles, and alternating with those which have preceded them, develop in the same way, and eventually are merged into an apparently single circle, as I have described in Tubularia.

Fig. 140*a*.

Fig. 140*b*.

There are, however, certain Hydroids, whose tentacles are unquestionably arranged in a spiral; for instance, those of Rhizogeton (p. 73, fig. 38) and Coryne (p. 78, fig. 42). But the spiral in this case has a different relation to the organization from

Fig. 140*a*. The planula stage of *Aurelia flavidula*. Per and Les. 100 diam. *m*, the mouth; *d*, the cavity of the body. The body is covered by vibratile cilia. — *Original*.

Fig. 140*b*. The primary, or bitentaculate stage of the Scyphostoma of *Aurelia flavidula*. Per and Les. 100 diam. *s*, the stem; *m*, the mouth; *d*, the digestive cavity; *t*, the tentacles. — *Original*.

what obtains in Protozoa; in the latter the *plane* of the axis is rolled in a spiral, whereas in the Zoöphytes the plane is fixed and the spiral winds around it.

Among all the other forms of Zoöphytes there are none, neither among *Polyps* nor *Starfishes*, of which it could be urged, with the faintest show of reason, that they are related to *Protozoa*. Hydra, and its congeners, although the best and most favorable instances of relationship to Protozoa, failing to hold what is claimed for them, the rest of the Zoöphytes are of course out of the question.

*Protozoa and Mollusca.* We will pass now to the consideration of the alleged relations of the Vorticellæ (one of the several groups among *Protozoa*) to *Mollusca*. As long ago as 1846, Van der Hœven, a Hollandish naturalist, suggested the relationship of the Vorticellæ to Mollusca, saying that "probably one day they will be ranked among Bryozoa." The latter is one of the lowest groups of Mollusca. Since that time others have repeated his suggestion. It is my task now to ascertain what amount of truth there may be in the Hollandish naturalist's prophecy. As I have already described and recapitulated (p. 160–176) the structure of the Protozoa, I need only refer to that group whilst I proceed to take in hand the Bryozoans.

The remarkable symmetry of the various constituents of their organization, which I have spoken of on a former occasion, (p. 195,) will be recalled to your minds by reference to these figures, (figs. 118, 119, 120, Pectinatella and Fredericella); and therefore I need not enter into any details in regard to that point.' What is most likely to attract your attention, in the comparison of the two groups, is the distinctness of the stomach ($st$) and intestine ($cl$ to $an$) of the Bryozoa from the rest of the body, and the equally free play allowed to the muscles, ($r$, $r^1$,) to say but a word in regard to the distinctness and especially assigned position of the nervous ganglion ($g$). When, therefore, we analyze the impression which we receive from the contemplation of these two groups, *e. g.*, Vorticellæ (fig. 95) and Bryozoa (figs. 118,

* The symmetry of the young Bryozoan, and the peculiarity of the type to

119, 120), it turns out that nothing but a certain general, external resemblance in form is the basis upon which their relationship has been claimed.*

*Protozoa and Articulata.* The fallacy of certain alleged tran-

which it belongs, become evident at an early period. This may be as well illustrated by its growth in the bud as by that in the egg. The bud commences as a mere internal, globular projection, (fig. 120, *bd*, and fig. 140*c*, *a*,) formed by the two walls (fig. 140*c*, *w*, *w*¹) of the general cavity. The first indication of a cavity appears in the form of a hollowing (*ca*) in the core of the bud. In process of time the bud elongates and doubles upon itself, (fig. 140*c*, *t* to *st*¹,) and, one end becoming broadened, a circle of feelers, (*t*,) fourteen in number in Fredericella, becomes apparent, at first as a scalloped edge, whilst the internal wall is differentiated into a general muscular layer (*w*², *w*³) and a group of distinct muscles (*r*, *r*¹, *r*².) like those of the adult. When the organs have developed to a certain degree, the walls of the stem opposite the two ends (at *t* and *an*) of the body of the embryo are perforated by a process of resorption, and the young Bryozoan is at liberty to protrude its head, and commence its first meal.

Fig. 140*c*.

* In all probability Van der Hœven was led to suggest the relationship of Vorticellæ to Bryozoa by the assertion of Ehrenberg that the former possesses a distinct intestine which doubles upon itself, somewhat in the way that it does in the latter. Subsequent naturalists, being led to disbelieve the truth of Ehrenberg's observations, have had far less reason than the Hollandish observer to claim such a position for the Vorticellæ.

Fig. 140*c*. The end of a branch of *Fredericella regina*, Leidy, MS., with two young budding. 100 diam. *ek*, the parchmenty sheath; *w*, the outer, and *w*¹, *w*², *w*³, the inner walls, the latter having a semi-muscular nature; *a*, *b*, the two walls of the incipient bud; *ca*, the first trace of a digestive cavity; *t*, the incipient tentacles of a far-advanced bud; *st*¹, the stomach; *an*, the posterior end of the intestine; *r*, *r*¹, the right and left retractor muscles of the head; *r*², the retractor muscle of the intestine, projecting over *st*¹ like an outer wall at *r*³, and forming a continuous layer over the head (at *r*⁴) in connection with *w*². — *Original.*

sitions between *Protozoa* and *Articulata* is the more difficult to detect because the examples brought forward from the latter group are so simple in their organization as to present a meagre basis of comparison. When you call to mind what has already been said in regard to the marked resemblance between all animals at a very early period of life, you can readily imagine how slight must be the means of distinguishing two diverse kinds which have just begun to lose their general resemblance, and to assume the peculiarities proper to the type to which each respectively belongs. This, in all probability, is the condition of the instance which I shall now lay before you. At first sight this creature, *Opalina*, (fig. 141,) strikes one as being most decidedly Infusorian. This arises from the fact that it is slightly one-sided, and covered with vibratile cilia. A closer inspection, however, does not reveal the presence of a mouth where one would expect to find it, that is, near the inward curvature at the narrower end, nor at any point. As for the vibratile cilia, they occur in a large number of animals beside the Protozoa, so that their presence alone cannot have any weight. The so-called contractile vesicle (*h*) of Opalina is described by the discoverer of this species to "have this peculiarity, that, at times, instead of undergoing a total contraction, it constricts itself from point to point in such a way as to transform itself into a series of rounded vesicles, disposed, one after the other, like the beads of a rosary." This is a characteristic of the pulsating vessels of a certain class of *worms;* and what renders the relationship the more certain is that those worms, which moreover are mostly parasitic and intestinal, have neither mouth nor intestine.*

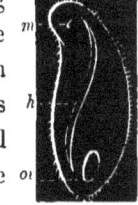

Fig. 141.

There are other species of Opalina, closely allied to this,

---

* See the description of Tænia, and the general remarks upon intestinal worms, on pages 79 to 84.

Fig. 141. *Opalina recurva.* Clap. 150 diam. A ciliated, infusorian-like worm. *m*, the hook-shaped body; *h*, the pulsating vessel; *ov*, the reproductive organ. — *From Claparède.*

which form a series of transitions of a most unmistakable
character, finally leading from the one before us to those which
have a decidedly *articulate* body. It is to be noted that the latter
as well as the former, and the species before us, have a so-called
"nucleus," *i. e.*, a reproductive organ, (*ov*,) and moreover the
peculiar hook-shaped body which is so prominent in some of the
jointed forms is also present (*m*) in the one which I have illustrated. In conclusion, therefore, it may be said that everything
tends to show that the Opalinas are members of the group of
Articulata, whilst, on the other hand, their relationship to Protozoa is based upon far-fetched resemblances.

*Zoöphyta and Articulata.* I have already spoken of the con-
tested position of certain worm-like animals, which had beer.
classed among the Zoöphytes by the earlier naturalists, and as I
gave such a full description of two of these forms, *e. g.*, *Caudina* (figs. 114, 115, 116) and *Bonellia* (figs. 126, 127), which
most closely approximate each other, from each side of the dividing line between Zoöphyta and Articulata, I shall only recur
to them again simply to mention the most characteristic features
which separate the one from the other. In the *Caudina*, (fig.
114,) it is the strict conformation to the Zoöphytic type, the systematic lateral repetitions of the organs, such as the aquiferous
circulatory tubes, ($aq$ to $aq^i$,) the nervous cords, (fig. 116, *n*,) and
the muscular bands (fig. 115, *ms*) along the sides of the body,
which distinguish it from all Articulata; and, on the other hand,
in *Bonellia*, (fig. 126,) it is not only the single ventral nervous
cord ($g, g^1, g^3$) and its peculiar loop ($g^2$) about the throat, but
also its series of numerous, successively repeated, longitudinally
arranged ganglionic enlargements, ($g^1$,) and the therefrom arising
transverse nervous threads, which give it an unquestionable claim
to be ranged in the type of Articulata; whilst a total absence of
anything like a lateral repetition of the organs offers no chance
whatever for a comparison in this respect with the Zoöphyta;
not even with the singularly worm-like *Caudina*.

*Bryozoa and Zoöphyta.* It is contended by eminent authority, even at the present day, that the Bryozoans belong to the

group of Zoöphyta rather than to Mollusca; but there are others, whose opinion is equally as worthy of attention, who claim for them a place among the latter. As frequently happens, the only basis for the assertion of their zoöphytic affinities is that of general resemblance, no more than you may see in these two figures, one of a *Polyp* (Metridium, fig. 106) and the other of a *Bryozoan* (Fredericella, fig. 120). If, now, we cover the head of each, the circle of feelers is concealed, and all trace of resemblance between the two vanishes. It is true that in both there is a free digestive cavity ($st$) or stomach, properly speaking, but in the Polyp it (fig. 106, $st$) terminates (at $p$) within the great body chamber, so that the food is left to the final assimilation by the general internal surface; whereas in the Bryozoan there is a complete intestinal canal, (fig. 120, $æ, st, cl^1$,) within which the digestion is carried on exclusively. But what forms a most decided separation is the difference in the relations of the surrounding parts. In the Polyp the organs are repeated like parallel lines along the sides of the body, and have a direct reference to the disposition of the feelers. This I need not describe in detail, as it has already been done (pp. 57 to 60, and in chap. x.) in former lectures. In the Bryozoan, (fig. 120,) on the contrary, there are no such lateral duplications of similar organs, but everything is devoted to a unity of purpose; there is the single retractor muscle ($r^2$) of the intestine, the single pair of right and left muscles ($r, r^1$) for withdrawing the head within the sheath, ($ek$,) and the unique nervous ganglion ($g$) placed exactly on the middle line of the head, all of which find no parallel in the Polyp.

## CHAPTER XV.

### THE MIMETIC FORMS OF DIVERSE TYPES OF ANIMALS.

My principal aim in the last lecture was to show, by comparison, the distinct lines of separation which exist between the *five grand divisions*, which I had stated were characterized, in the several groups, by certain relations, through means of which they might be recognized, and distinguished, the one from the other. Thus the Protozoa are characterized by the relation of their organization to a *spiral* (p. 175). The Zoöphyta repeat their organization in *parallel lines, along the length of the body* (p. 177). The Mollusca are based upon a *uniformity* of organization; they are *monomerous* (p. 195). The Articulata *repeat their parts from point to point* along the body, so that it appears jointed from one end to the other (p. 214). And the Vertebrata have an upper and a lower region, separated from each other by the vertebral axis (pp. 126, 231). I then referred to the so-called transitions from one division to another, and showed that they could not hold out, that they were illusory.

Now the reason why it has been, and is, even at the present day, so difficult, in some instances, to discover the proper position of certain animals, is twofold. In the first place, it often happens that a newly discovered creature is not in its adult state, and in consequence of a lack of development its characteristics have not been marked out.

You will recollect that I have said that all animals in their earliest stages of growth possess certain characters in common. This is the condition in which animals, belonging to one grand division, may oftentimes be mistaken for those of another division. How much, for instance, this figure (fig. 145) of a young Starfish, or this of a young Molluscan (fig. 149), resembles a worm. Place them side by side with this embryo (fig. 148) of a genuine worm, and no one could say positively, if they were new to him, which is not a worm. Until it passes beyond this embryonic condition, it is oftentimes extremely difficult to determine in which of the five grand divisions an animal belongs.

This difficulty arises from the fact that the predominant idea in all animals is *bilaterality;* and for the reason that during the early stages of growth this idea is expressed with a degree of exclusion which obscures the subordinate idea, which finally is *superimposed* upon the *predominant one.*

Let us follow out the development of this idea among the examples we have here, and see how it addresses itself to our eyes. In these two figures (figs. 142, 143) of young Trepangs, you will notice that the arrangement of the visible parts of the organization is in reference to a line which may be drawn lengthwise through the middle of the body; that is, the organs are either immediately upon the axial line, as in this figure, (fig. 142,) or situated symmetrically right and left of it, as represented here (fig. 143). This feature is the most prominent one; and notwithstanding there is the beginning of a lateral repetition of parts in one of the figures, (fig. 143,) it is not so decided, nor so far developed, as to

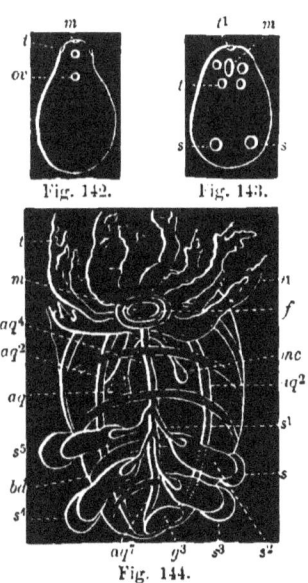

Fig. 142. Fig. 143.

Fig. 144.

Fig. 142. *Holothuria tremula.* Gunn. Natural size about $\frac{1}{25}$ of an inch long. A view of the back of a very young Trepang. *m*, the future mouth; *t*, feelers beginning to develop; *or*, the madreporic body, lying beneath the skin. — *From Koren and Danielssen.*

Fig. 143. The same as fig. 142, but a little older. A view of the lower or ventral side. *m*, the mouth developing; *t, t¹*, the incipient feelers; *s*, the first pair of sucker-like feet. — *From Koren and Danielssen.*

Fig. 144. *Holothuria tremula.* Gunn. The same as figs. 142, 143, at a far

enable one to determine whether the embryo, *per se*, is one of the Zoöphytic types, or is one of those worms (Tape-worms, &c.) which bear a crown of hooks or suckers about the mouth, or is a member of that group of Trepang-like worms to which Bonellia (p. 216, fig. 126) belongs, and of which some possess a circle of feelers at the end of the head. In fact, I have already stated that some good naturalists persist to this day in classing this latter group, the Sipunculacea, with the Holothuria (Trepangs).

If, however, we trace the development of these embryos, as did the Norwegian naturalists from whose work these figures are copied, we shall find that eventually, as the other parts of the organization become prominent, as they appear in this figure, (fig. 144,) they demonstrate by their arrangement a decided lengthwise lateral repetition (see the description of the figure at the bottom of the page).

In these other two figures, (figs. 145, 146,) which represent the two opposite sides of the same individual, (an embryo Starfish,) one of them (fig. 145) has all the characters of a jointed worm, with its joints, as it were, strung on a longitudinal axis; nor does the other (fig. 146) present a much less articulate form than the former; but yet for those who are familiar with the

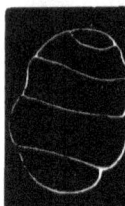

Fig. 145.

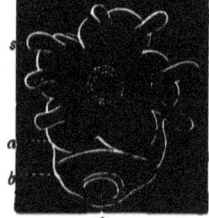

Fig. 146.

more advanced period of development. Considerably magnified. (See chap. XVII. § ZOÖPHYTA, in the section on *Holothuria*.) A view of the lower side. *t*, *t¹*, the feelers; *m*, the mouth; *f*, the buccal ring; *aq*, one of the sack-form appendages, *Polian vesicle*, of the circular aquiferous canal around the mouth; *mc*, the madreporic canal; $aq^1$, $aq^7$, the median longitudinal aquiferous canal; $aq^2$, $aq^3$, the right and left inferior aquiferous canals; *s*, $s^3$, $s^4$, $s^5$, the disciform ends of the sucker-like feet ($s^2$) which arise from the lower median line of the body; $s^1$, the aquiferous tube by which the feet are distended with fluid from the median vessel ($aq^4$); *bd*, the transverse muscular bands; the five longitudinal muscular bands occupy the same place as the longitudinal aquiferous canals, ($aq^2$, $aq^4$,) each one

young of this group, there is sufficient of the starfish physiognomy about its anterior end, to enable them to refer it to its proper place among the Zoöphytes.

Here, again, in these two figures (figs. 147, 148) of a young worm, the younger one (fig. 147) shows scarcely 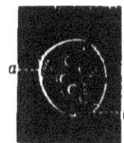 the trace of articulation, whilst its bilaterality is prominently set forth by the pair of eyes, and the two right and left groups of vibratile cilia (*a*). In the older one (fig. 148) it is clear that the jointed structure of the body is superimposed upon a bilateral basis; for it is still quite as evident as in the younger one, that not only the eyes, but also the groups of vibratile cilia (*a*) are arranged in reference to a right and a left, notwithstanding the partial obscuration of this feature by the prominency of the articulation.

Fig. 147.

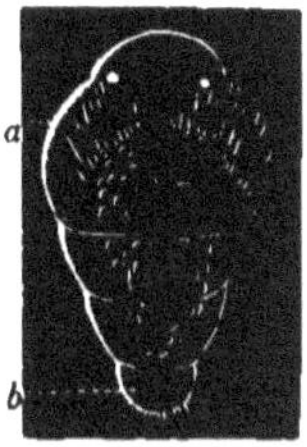

Fig. 148.

Finally I will recall to your minds the young worm-like Molluscan, (fig. 149,) simply to show how little it resembles its adult state, as represented in this figure, (fig. 150,) and yet how strictly it conforms to the bilateral type.

If, now, we turn to the lowest forms of each grand division, we shall find this idea carried out in *another guise*. This will

---

of the latter lying in a furrow along the middle of one of the former; $g^3$, the cloaca, a saccular enlargement at the posterior end of the intestinal canal.— *From Koren and Danielssen.*

Fig. 145. A view of the back of a worm-like embryonic stage of a *Starfish*. Magnified about 40 diameters. — *From J. Müller.*

Fig. 146. A view of the front side of the same as in fig. 145. *s*, the sucker-like feet in an incipient stage of growth; *a*, the end of one of the five short, blunt, rounded arms, from which *s* arise; *b, c*, the joint-like divisions of the body. — *From J. Müller.*

Fig. 147. *Protula elegans.* M. Edw. Highly magnified. A view of the back of a young marine worm, in its earliest stage, at the moment of its birth. *a*, the groups of vibratile cilia at the anterior end; *b*, the posterior end; the eyes are two dots just in front of *a*. — *From M. Edwards.*

Fig. 148. The same as fig. 147, in a more advanced stage of growth. *a, b*, as in fig. 147. — *From M. Edwards.*

explain the second branch of the difficulty. You 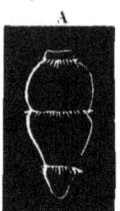 will recollect that I have said (page 158) that the *higher animals* of a *group* pass through certain forms in their development which correspond to the *permanent adult* condition of the *lower animals* of that same group; e. g., the Rabbit passes from that degree of or- 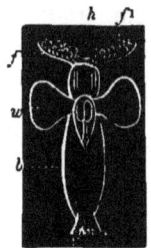 ganic simplicity which corresponds to that of a fish, to a higher state which resembles that of a reptile, and then onward to that phase which is typical of the organization of a bird, and finally it assumes its adult condition. A Butterfly, one of the highest of the group of Articulata, in its youngest stages is a grub or worm, so-called, whose numerously jointed body and frequently repeated internal organs recall the many-jointed, true worms, such as the *earth-worm*, and marine worms, (p. 80, fig. 43,) but more strictly correspond to those of the *earwigs*, and the *thousand-legs*, or *ringed worms*, so common under sticks and stones on the ground. This idea was first put forth by Von Baer, thirty-six years ago, in 1828, (Über Entwickelungsgeschichte, &c., Theile I. p. 230,) and, taking the cue from him, other observers have traced its prevalence throughout the animal kingdom, not only among the living, but also through the numerous groups of extinct fossil animals of past ages.

From this you may judge that all the members of a group do not come to the same degree of perfection, but that some remain in what corresponds to the embryonic stages of the higher animals of that group. Now from this you may also infer that we would have the same difficulty with these permanently embry-

Fig. 149. *Pneumodermon violaceum.* D. Orb. Natural size, 0.15′′′, i. e., $\frac{15}{100}$ of a line long. A worm-like stage of development. A, the head; B, the tail. The body is encircled by three groups of vibratile cilia. — *From Gegenbaur.*

Fig. 150. *Pneumodermon violaceum.* D. Orb. 3 diam. An adult Pteropodous Molluscan which swims in the open sea. $h$, the head; $b$, the main part of the body; $c$, the tail; $f$, the pair of feelers; $f^1$, the suckers of $f$; $w$, the fins which are used like wings. — *From Woodward.*

onic forms in determining their position, as we noticed in regard to the young trepang, starfish, worm, or molluscan; for example, one of the lowest fossil forms of the group to which the Zoöphytic Trepang belongs, a *Hemicosmites* (p. 128, fig. 66) has been mistaken for one of the Mollusca, a *Chelyosoma*, whose body is covered by similar, many-sided, pavement-like scales. Again, certain of the group of Articulata, for instance, some of the Planarians, (like fig. 47, p. 92,) e. g., *Eolidiceros*, have a remarkable resemblance in form to certain of the so-called naked Mollusca, namely, *Janus*, and also exhibit the same habits, in the manner of walking, and also in crawling back-downwards along the surface of the water. It is not a little curious, too, that *Eolidiceros* and *Janus* should resemble each other in regard to their nervous system.*

Among Mollusca, the *Bryozoa*, especially those which have a simple circle of tentacles around the mouth, have been mistaken for *Polyps*. This point I have already discussed, when comparing the *Bryozoa* with *Polyps*, in a previous lecture (p. 247).

Even among Vertebrates, there are those, for instance, certain kinds of Lamprey-eels, (*Myxine*,) which might at first sight be mistaken for worms; and in fact they have been described as such by Linnæus; and others, such as the *Lancelet*, (Amphioxus,) (fig. 133, p. 226,) have been wrongly classified with *sea-slugs*, a group of naked Mollusca. This was owing to the undeveloped state of the vertebral axis of the *Lancelet*, and the otherwise singular resemblance of this fish to certain Mollusca (*Firolidæ* and the like) which swim rapidly in the open sea.

* We owe to Blanchard (Voyage en Sicile, vol. III.) the first decided and conclusive proof that the Planarians are true Articulates, by his demonstration of the successive ganglionic repetitions along the nervous threads, at the right and left sides of the mid-line of the body of a large Planarian (*Polycladus Gayi*. Blanch.).

## CHAPTER XVI.

#### THE TRANSITIONS AMONG THE SUBORDINATE TYPES OF THE FIVE GRAND DIVISIONS.

SINCE, therefore, there are no transitions from one grand division to another, it is very natural to infer that it is among the groups of *each particular division* that we shall find the passages from one form to another. This inference corresponds, too, with the strongest tendencies of belief among the majority of naturalists in regard to the relations of animals; for although there are many who more or less incline to see transitions from one grand division to another, yet it is very seldom that any one makes a positive assertion to this effect; whereas, in regard to the transitions from *one class to another*, e. g., from Fishes (through Lepidosiren, fig. 169) to Reptiles, or from Reptiles (through Pterodactylus, fig. 186) to Birds, (through Archeopteryx,) we have the strongest expressions and most positive assertions that the one has no definite boundary which separates it from the other.

If I now place before you certain pictures of animals, one from each *class* of the five grand divisions of the animal kingdom, such as the advocates of the fixity of the subordinate types would select, you will no doubt be inclined to say, "How very clear the proof is; these must be distinct types." I have arranged in this table the names of the classes of each grand division in groups of relationship, so that you may inform yourselves as to the alliance of one class with another, according to their degree of proximity.

## VERTEBRATA.

*Mammals* (fig. 151.)
*Birds* (fig. 152).
*Reptiles* (fig. 153).
*Fishes* (fig. 154).

## MOLLUSCA.

*Cephalopoda* (fig. 155).
*Gasteropoda* (fig. 156).
*Acephala* (fig. 157).
*Bryozoa* (fig. 158).

## ARTICULATA.

*Insecta* (fig. 159).
*Arachnida* (fig. 160).
*Crustacea* (fig. 161).
*Vermes* (fig. 162).

## ZOÖPHYTA.

*Echinodermata* (fig. 163).
*Acalephæ* (fig. 164).
*Polypi* (fig. 165).

## PROTOZOA.

*Ciliata* (fig. 166).
*Suctoria* (fig. 167).
*Rhizopoda* (fig. 168).

Description of figs. 151 to 168, on pages 257 to 261:

VERTEBRATA.

Fig. 151. See p. 118, fig. 51, for the description of the lettering.
Fig. 152. *Dendroica virens.* Baird. Black-throated green warbler. ½ natural size. — *From Audubon.*
Fig. 153. *Scleroporus consobrinus.* B. and G. Natural size. A lizard, from our southwestern territory. — *From Baird & Girard.*
Fig. 154. *Argyreus atronasus.* Heck. Natural size. The common *brook minnow.* — *From Storer.*

MOLLUSCA.

Fig. 155. *Loligopsis illecebrosa.* ⅔ natural size. The common *squid* of this coast. For the description see p. 206, fig. 124.
Fig. 156. *Helix albolabris.* The common *snail.* 2 diam. For the description see p. 203, fig. 122.
Fig. 157. *Ostrea Virginica.* L. The common *oyster.* For the description see p. 200, fig. 121.
Fig. 158. *Fredericella regina.* Leidy, MSS. One of the fresh-water Bryozoa. For the description see p. 198, fig. 120.

ARTICULATA.

Fig. 159. *Sphinx Ligustri.* L. Natural size. The *hawk privet-moth.* For the description of the lettering see p. 119, fig. 53.
Fig. 160. *Epeira trifolium.* Hentz. Natural size. A common *garden spider.* For description of lettering see p. 220, fig. 129.
Fig. 161. *Cyclops quadricornis.* 50 diam. A minute, shrimp-like Crustacean. For description see p. 220, fig. 128.
Fig. 162. *Myrianida fasciata.* M. Edw. 2 diam. A marine worm. For the description of the lettering see p. 80, fig. 43.

ZOÖPHYTA.

Fig. 163. *Psolus phantapus.* Strlldt. ⅓ natural size. A *trepang* from this coast. For description of lettering see p. 192, fig. 117.
Fig. 164. *Tubularia indivisa.* L. Natural size. A *Hydroid Acaleph* from Boston harbor. For description see p. 238, fig. 136.
Fig. 165. *Metridium marginatum.* M. Edw. ½ natural size. A *Polyp* from Boston harbor. For description see p. 57, fig. 28.

PROTOZOA.

Fig. 166. *Epistylis flavicans.* Ehr. 250 diam. One of the highest of the group of *Ciliata.* For description see p. 161, fig. 95.
Fig. 167. *Podophrya Cyclopum.* Clap. 300–350 diam. A pedicellate infusorian belonging to the group *Suctoria.* For description see p. 51, fig. 25.
Fig. 168. *Amœba diffluens.* Ehr. 100 diam. An *Amœba* or *Protean animalcule*, belonging to the group of *Rhizopoda.* For description see p. 9, fig. 1.

OF ANIMALS.   257

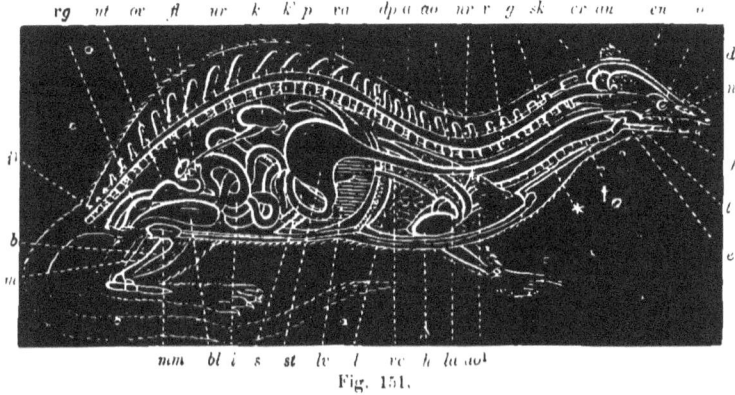
Fig. 151.

Fig. 152.

Fig. 153.

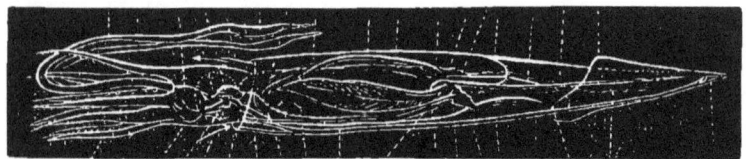

Fig. 155.

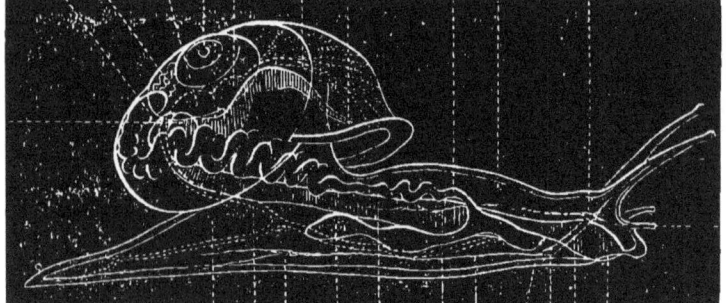

Fig. 156.

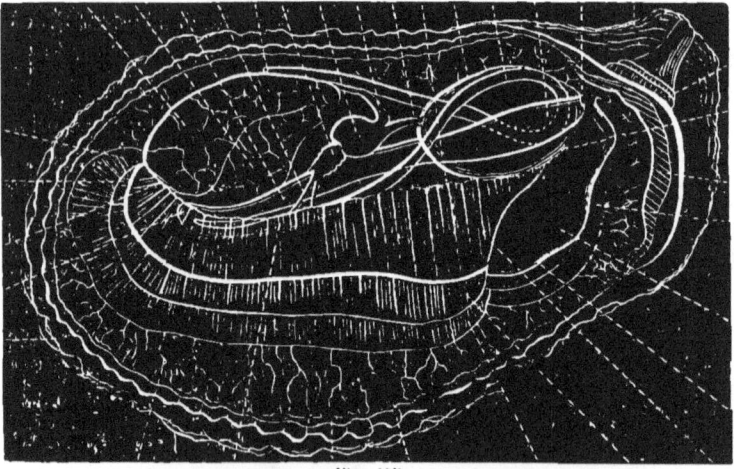

Fig. 157.

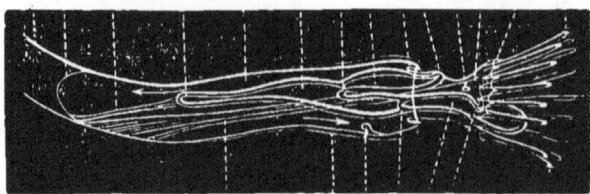

Fig. 158.

OF ANIMALS. 259

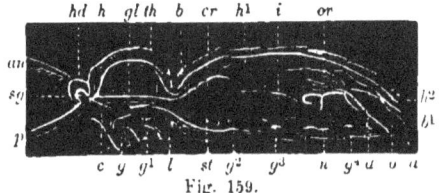

Fig. 159.

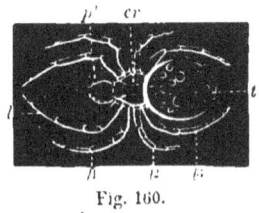

Fig. 160.

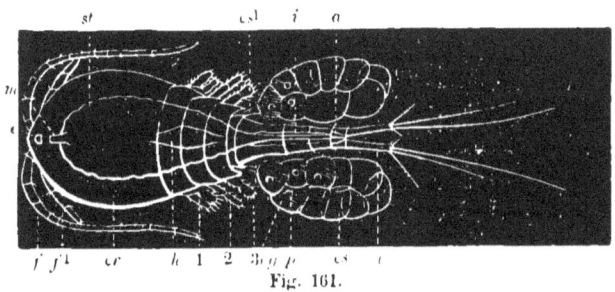

Fig. 161.

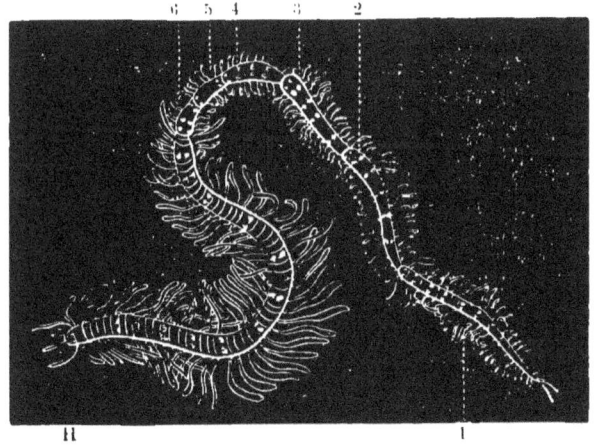

Fig. 162.

260 THE CLASSIFICATION

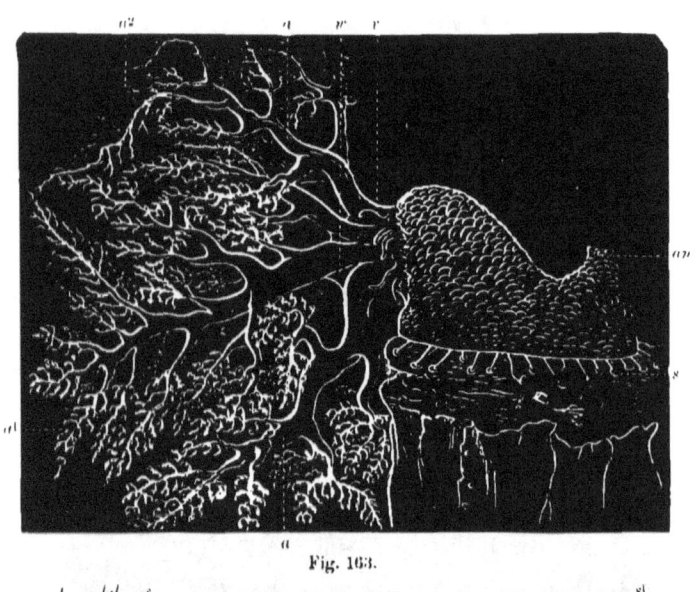

Fig. 163.

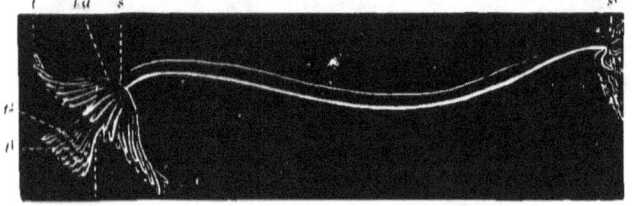

Fig. 164.

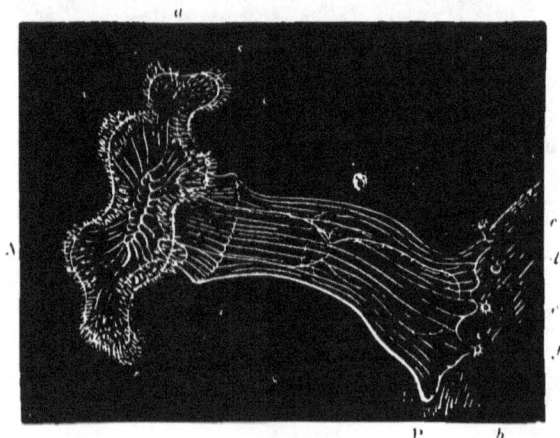

Fig. 165.

OF ANIMALS. 261

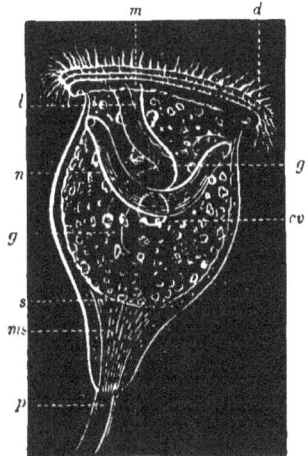

Fig. 166.

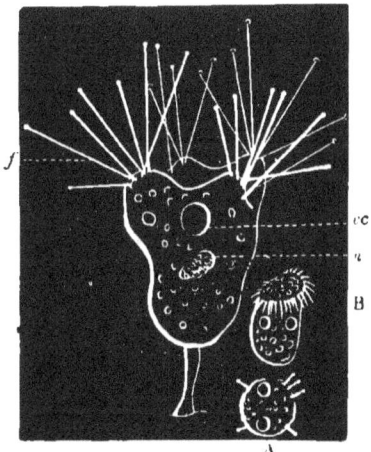

Fig. 167.

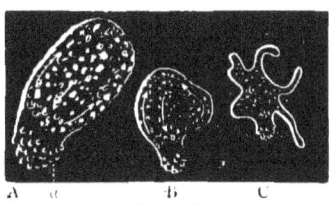

Fig. 168.

Now it would be impossible for me to show all the details of transition among these various groups, from one into another, because, in the first place, time would not allow it, and secondly, it could not be done except by the actual inspection of specimens; and even then the demonstration would have to be illustrated in a superficial manner. These transitions, in the majority of cases, are such as are addressed to the eye and thought of the well-studied observer of nature; and when they can be illustrated, they afford a view of only the more general relations. It is on this account that, in a course of lectures like these, many things are to be taken for granted upon the assertion of the lecturer, whilst occasionally he is able to bring forward more or less of his proofs in detail. In the latter case, the presentation of facts involves a form of special pleading which can hardly be addressed to an audience which is not directly and practically interested in the subject. The dryness of the details are beyond the patience of those who, at most, look forward to the final exhibition of the more general results.

It would be extremely onerous, for instance, to attempt to show in what points certain Fishes and certain Reptiles cannot be distinguished from one another, and in what points they essentially or manifestly differ. We can only say, for instance, that certain kinds of Catfish or Horned-pouts have an organization so much like that of certain Salamanders and frog-like Reptiles, that naturalists class the two together in one group; and that not only are they related in this most general way, but there are among them certain forms which cannot be definitely and unequivocally proved to be either Fish or Reptile; and on this account naturalists are all the more inclined to look upon two groups which have no definite boundary line as merging into one another. Here is an instance in illustration of this relation.

This creature (fig. 169) has a fish-like form, but so have the tadpoles of Salamanders, and certain adult Salamanders. Its body is covered with scales, whose structure, however, is not like that of the scales of ordinary fishes, but rather of the Lizards (fig. 153). It has *internal* gills like those of a fish, but a part

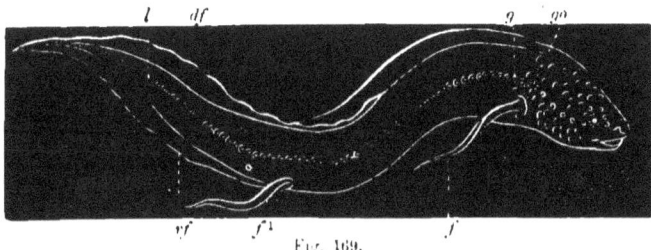

Fig. 169.

of them do not perform the functions of such organs, and are the remains of gills of a younger state; just such a structure as we may see in the remains of gill-like bodies on the gill-arches in certain salamanders, *e. g.*, Siren. In its younger stages of growth it has tufts of *external* gills on each side of the head, like those of the tadpole of frogs and toads, and the permanent ones of certain adult salamanders, such as *Siren, Axolotl,* and *Menobranchus;* but on the other hand they equally resemble those of the embryos of some of the sharks and skates. The heart is not a double cavity as in the bony fishes, but the posterior half is divided into two; yet as the partition in this case is a mere net-work, the division is not a complete one, but of that sort which obtains among the so-called *Reptilian fishes;* and so I might go on to enumerate every character which this singular genus exhibits, without giving one which does not find its counterpart either in fishes or reptiles. There is one character, however, which has determined some naturalists to class the Lepidosiren among fishes, and that is, *it has no nostrils,* and on this account Owen decided that it is rather a fish than a reptile; but in specimens which Bischoff examined, *the nostrils are present;* and therefore the preponderance passes over to the side of reptiles, if Bischoff's opinion is alone to be taken into account. Between the two authorities, however, the difference is so slight as to amount, I suspect, to an individual disparity;

Fig. 169. *Lepidosiren annectens.* Owen. ¼ natural size. From the River Gambia, Africa. *go,* the gill-opening; *g,* the remains of the external tuft of gills of the youngest stage of growth; *f,* the anterior, and *f¹* the posterior, flattened, lash-like limbs; *df,* the dorsal fin-like crest; *vf,* the ventral fin-like keel; *l,* the row of scales in the trend of the so-called *lateral lymphatic canal;* excepting on the head, the rest of the scales are omitted.— *Original, from a specimen kindly lent to me by Prof. Jeffries Wyman.*

an amount of which no definite estimate could be made that would be sufficient to separate the animal in question either from fishes on the one hand or from reptiles on the other.

Now this is the way in which naturalists proceed, step by step, balancing, through all the intimate details of an organization, the several features which are opened to their eyes by the help of the dissecting knife and the microscope. Of course you must see, even from this slight sketch, that it would be a work of weeks, and even months, if I were to undertake to go through even the principal groups of the animal kingdom, to say nothing of the lesser communities of relations, such as exist among specific characters.

The case of the Lepidosiren is one among many that are just as remarkable and indeterminable, if one wishes to make definite, sharp boundary lines, like checker-board work, between all the various groups of animals. But the relations of Nature are not of this kind; the square, rule, and compass and the checkerboard formulæ are but the expressions of the limitations of human comprehension among the untutored; but as we advance in the knowledge of our own minds, and by degrees take in the immense breadth of intellect with which man is endowed, we gradually come to a better understanding of the workings of that greater mind which originally conceived, and gave us birth.

We have already had some glimpses of the *order of things* which reigns among the animal groups; we have seen that the *principle of life* is the expression of a grand idea which shines forth in all Organic Nature; we have seen that from and upon this idea is erected the idea of *form* in a *twofold* relation, that of the animal physiognomy on the one hand, and that of the plant on the other; but yet related in such a way that a conception of their difference would seem to be beyond the power of the finite human mind, and we fain would come to the conclusion that the difference is but one of *degree*.

For the animal kingdom we have traced an idea of progression which is expressed in several ways; and these ways are seen to refer more or less to each other, as if they were but the collateral branches of one main line. At the outset the principle of life takes on an outward form in the *bipolarity* of the egg. This principle is, as it were, set against itself; the opposing poles are antagonistic only in one sense, though; they cannot exist the one without the other, any more than can the positive and negative poles of the electrical sphere which wraps this globe in an invisible shell. The idea of a balance between these upper and lower poles seems, as a natural consequence, to necessitate the consideration of lateral relations; and these we find progressively elaborated and set forth in the idea of *bilaterality*. Everywhere we see the idea of *progression*, from a lower to a higher, manifested in various relations. The ideal relation of progress is that which is the most general.

And now when we come to the consideration of the characteristics of the five grand divisions, we seem to have lighted upon certain ideas which have no relation whatever to the fundamental idea of the *typical animal;* it seems at first glance as if *five new seals* were impressed upon the *original stamp*, and obscured or blotted out its distinguishing features; but I am inclined to look upon the matter in another light. You will recollect what I told you some time ago (p. 85) in regard to the tendency of the egg, especially of the lower animals, to divide into two or more, and that in this way the so-called monstrosities among higher animals were formed. You will call to mind, too, that I pointed out the prevalence of the *self-dividing* process of reproduction among the *lower* forms (chap. III. and p. 110). Among the Protozoa (Infusoria) this process takes place in such a way as to not only divide the animal transversely, but also lengthwise; that is, the creature not only repeats itself by crosswise division, but also by lengthwise or longitudinal repetition.

Now how do we see this idea carried out in the groups above Protozoa? In the Zoöphyte division we have taken note (p. 61) of the longitudinal splitting of the Polyps, especially of the compound forms, the branching corals. Among the single Polyps this splitting totally divides the individual into two; but in the branching corals the splitting only divides the head and stomach, and thus a many-headed creature is formed. The tendency would seem to be to separate the laterally repeated parts, as so many individuals, from each other.

We may observe everywhere in the animal kingdom that a manifold repetition of parts, whether longitudinally as in Polyps, or transversely as in Worms, is attended with an indeterminate relation of individuals: in some cases individuality is distinctly marked, whereas in other instances, and under various forms, the individual is more or less merged into the repeated parts.

Now in the Articulate animal group, where the repetition of parts is from point to point, along a line, the body is divided into joints, and, as if to exemplify the individual character of these joints, among the lowest intestinal worms, each joint at the period of transverse division is separated from its fellows, and plays, for a time, an independent part (p. 84). Among the aquatic worms I have already shown how a number of joints divide off together and form an individual (p. 80). Nowhere do we find in this grand group a longitudinal self-division, but always, in conformity with the type, a transverse one.

The Mollusca fully exemplify their type, as one which has a unity of conformation, with neither lateral nor longitudinal repetition of parts; for they do not possess the faculty of self-dividing. I mean, of course, that they do not, as a rule, take this method to reproduce themselves. It is true that some of the lower groups are compound; but they become so by a process which is purely a matter of budding, (p. 243, note,) and not of self-division, properly speaking. The instances of self-division or of fissigemmation, which I have pointed out as occurring, not only in Mollusca and Articulata, but even in Vertebrates, when discussing the matter of monstrosities, (p. 85,) were then given merely to show how universal an idea may be in its relations, although its practical and normal effects are confined to a more or less circumscribed circle.

Even the idea of *bilaterality* is involved with that of *reproduction*, and the manner in which this relation manifests itself, shows that the idea of a right and a left may possibly be the expression of *two individuals* united *side by side*, instead of the duplication of the organs of one individual. I have already referred (p. 85) to the remarkable investigations of Lereboullet upon the embryology of the Fish. Without stopping now to go into the details

of the subject, I will merely state some of the general results of
these investigations, as far as they have a direct bearing upon the
point which we are now discussing. Lereboullet discovered that
not only will an egg divide into two distinct individuals at the
outset, but also, in some cases, the two individuals are formed
so closely side by side that the right side of one and the left side
of the other become more or less united, and form, at the point
of junction, a single trunk. Sometimes this occurred at the tail,
sometimes at the middle, or at the head; and the manner in
which the united parts are joined to each other shows clearly
and unmistakably that the right side of an animal is not necessarily a different part of the body from, and in direct opposition
to the left side; for we see that the one may take the place of
the other; and we might say without exaggeration that they are
mutually interchangeable. This could not possibly happen, if
the two halves were not positively identical with each other. In
this sense, then, the body consists of *two individuals acting as
one;* it is a *duality.*

From this point of view, then, I think we may see that there
is a certain form of connection between the ideal relations of
*bipolarity, bilaterality,* and the *type of division,* on the one hand,
and, on the other hand, the *reproductive process,* which is the
basis of relation by *consanguinity. It is, then, through this process, that, from the apparently purely ideal relations of all animals
to each other, the transition is made to those relations or ties of
consanguinity which bind certain groups, as it were, in one family.*
We cannot say where the ideal relation meets or interlocks with
the relation by blood; the point of junction seems to vary in
the different groups. Among the lower forms of animals the
ideal relations of bipolarity and bilaterality are the most prominent, and the division type is but faintly shadowed forth; but,
as we rise in the scale, the latter becomes more decided, and
sometimes so strongly presents itself as to nearly obscure bilaterality. This is most notable among Starfishes and Jelly-fishes;
and it has misled some into the belief that *radiation* is the predominant feature to which all other characters are subordinate
in these animals.

But I have demonstrated (p. 177) that this so-called radiation is merely a lateral repetition of parts, and I have also shown, that, among the highest forms of these Zoöphytes, this repetition is less and less frequent, and that bilaterality is consequently more clear to the view.

Now, as we see the two ideas, represented by bilaterality on the one hand, and the type of a grand division on the other, mutually and alternately intensified, and bearing such a variety of degrees of the prominence of the one or the other; so may we also detect a similar relation between these two ideas and the characteristic features which blood relationship transmits from family to family. Will any one undertake to show how *bilaterality* is propagated from parent to offspring, if it is not an inheritance by blood; but if it is an inheritance by blood, at what point, then, does it cease to be altogether *ideal* in relation, and begin to be hereditary? I think it would be just as difficult to answer this question, as it would be to show where the ideal relation of the Lepidosiren to the Fish or the Reptile begins, and where the relation by consanguinity terminates.

I think I can illustrate the complicated interchanges of these ideas to the fullest extent by explaining the nature of the relations of the animals which are arranged here. It is a group of *Articulate animals*. Judging from their *forms* alone, you might conclude that there are several family types represented; — for instance, the worms seem to be illustrated in one place (*Lingua-tula*, fig. 170 (A); *Dendrocœlum*, fig. 182 (*M*)), and the spiders (fig. 176 (G)) in another. Others you might take for caterpillars (*Tardigrada*, fig. 178 (I)), and some for shrimps (*Squamella*, fig. 184 (O), and *Cyclops*, fig. 185 (P)). Now I could not say positively whether you would be right or wrong in some instances; for the very reason that several of these animals are related in *intermediate ways* to more than one family, and in a very complicated form of relation. On this account I will undertake to point out but a few of the features which characterize them; and in order that I may do this in the simplest, systematic manner, I have jotted down a tabulated arrangement of characters, which I will endeavor to illustrate by these figures.

# CERTAIN GROUPS OF ARTICULATA.

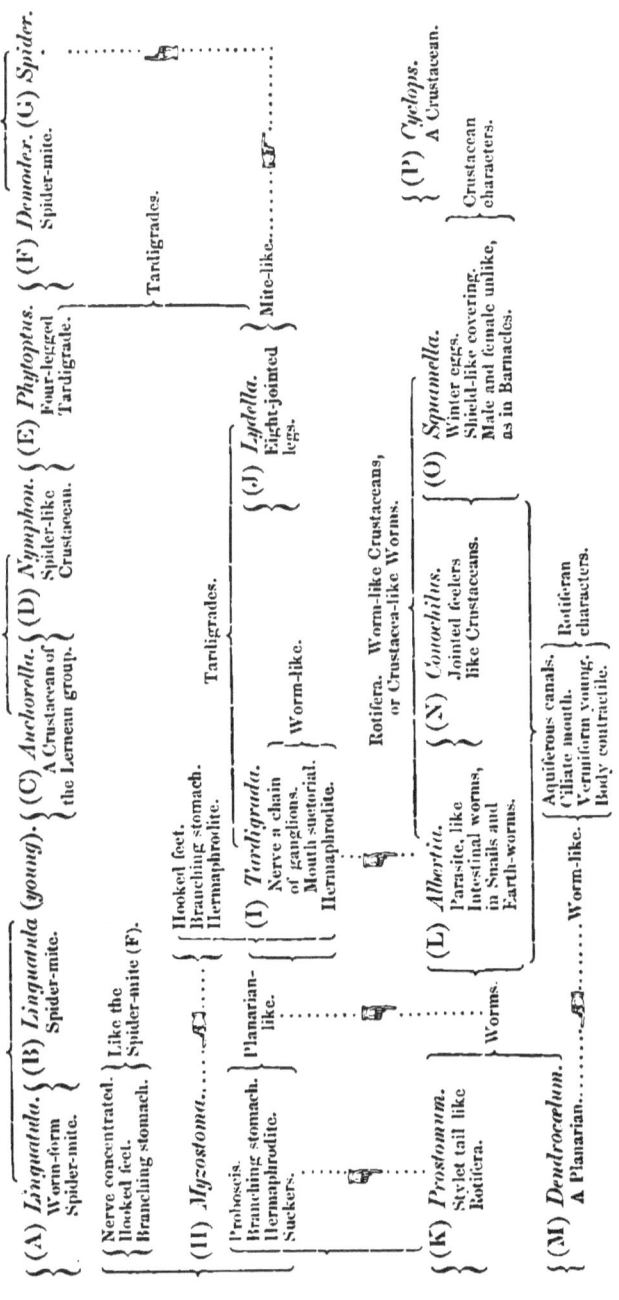

270   THE INTERCHANGEABLE RELATIONS OF

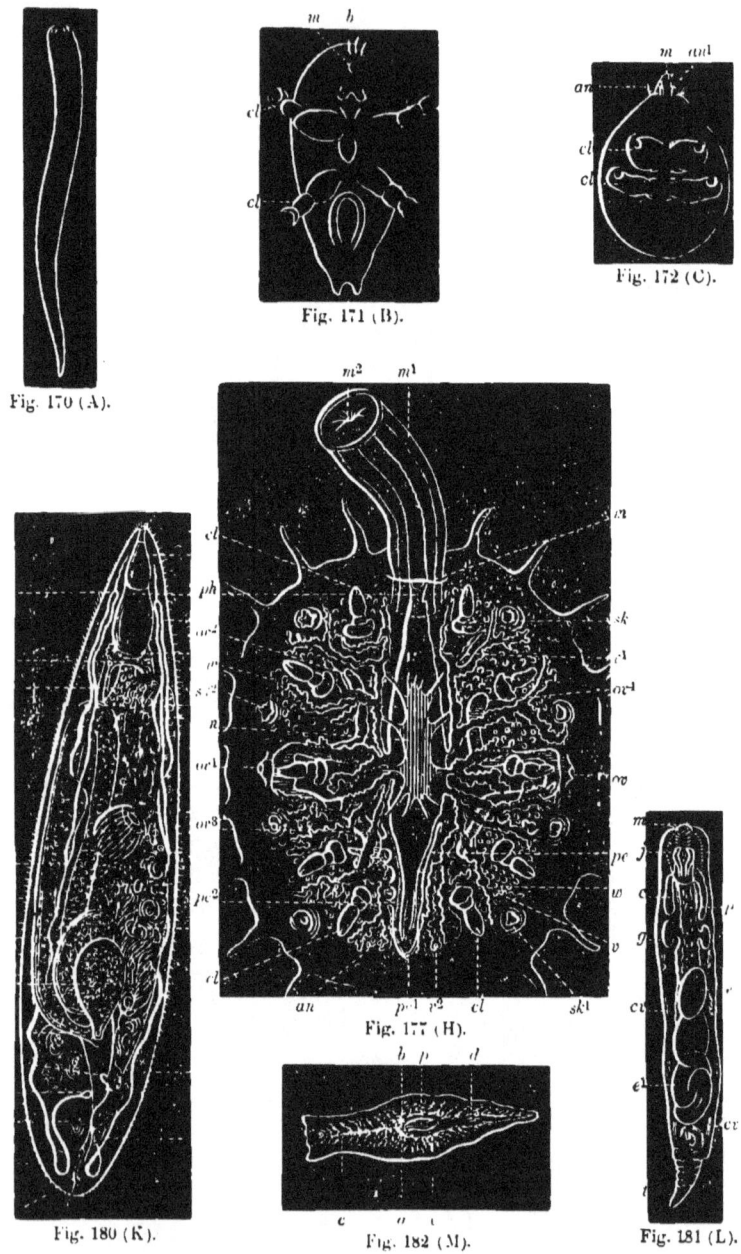

Fig. 170 (A).
Fig. 171 (B).
Fig. 172 (C).
Fig. 177 (H).
Fig. 180 (K).
Fig. 182 (M).
Fig. 181 (L).

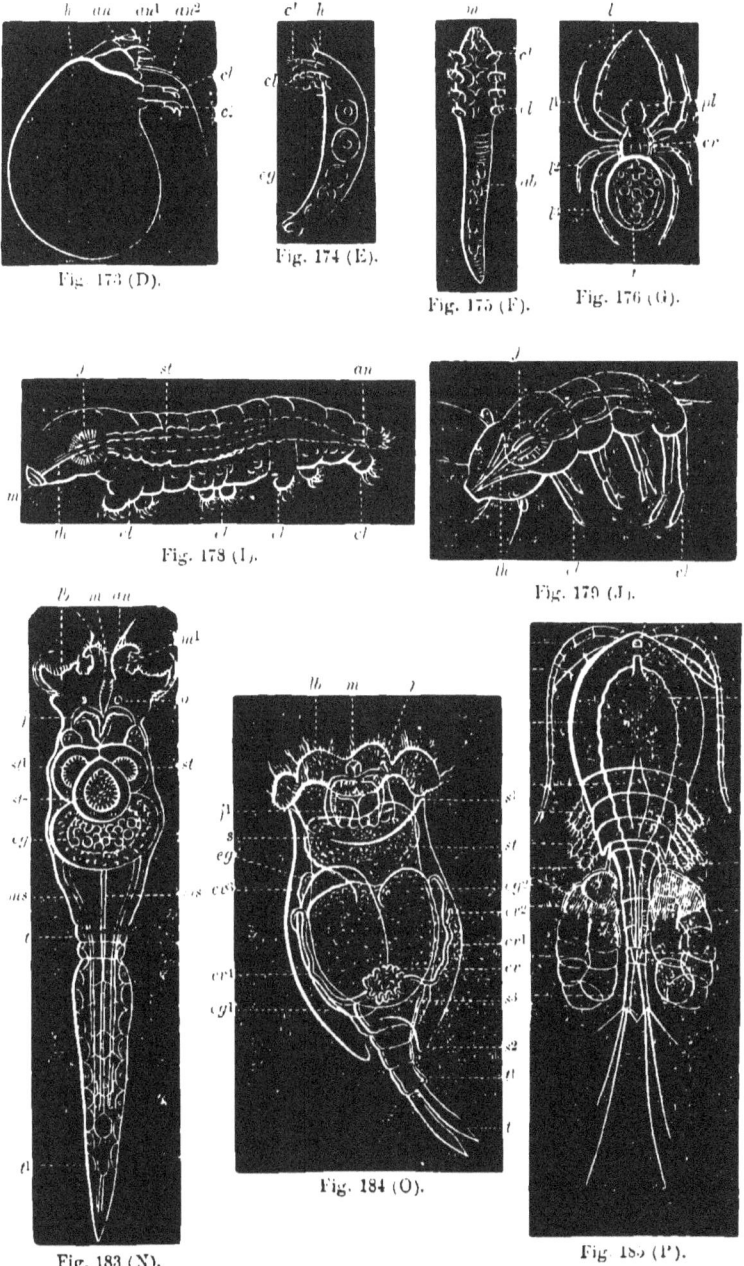

Fig. 173 (D). Fig. 174 (E). Fig. 175 (F). Fig. 176 (G).
Fig. 178 (I). Fig. 179 (J).
Fig. 183 (N). Fig. 184 (O). Fig. 185 (P).

This worm-shaped figure, (A) *Linguatula*, represents an animal which was at one time classed among the intestinal worms,

Fig. 170 (A). *Linguatula proboscidea*. Rud. About half natural size. At the broad end, which is the head, are a few recurved hooks. — *From Quatrefages*.

Fig. 171 (B). The young of fig. 170, just hatched. Highly magnified. *m*, the mouth; *b*, the bristles which are protruded from the mouth, and are used to pierce the tissues of animals; *cl*, the two pairs of hooked feet. — *From Van Beneden*.

Fig. 172 (C). *Anchorella uncinata*. Müll. Natural size about $\frac{1}{30}$ of an inch long. The male. *m*, the mouth; *an*, $an^1$, the two pairs of feelers; *cl*, the two pairs of hooked feet. — *From Nordman*.

Fig. 173 (D). *Nymphon grossipes*. Magnified considerably. A very young individual, seen in profile. *h*, the head; *an*, the pincer-like claws; $an^1$, the hooks of *an*; $an^2$, one of the feelers; *cl*, the hooked feet. — *From Kroyer*.

Fig. 174 (E). *Phytoptus*, sp. ? Dujardin. 150 diam. One of the Tardigrada with only two pairs of legs (*cl*); *h*, the head; *cl*, the legs; *eg*, the eggs. — *From Dujardin*.

Fig. 175 (F). *Demodex folliculorum*. 150 diam. A view of the lower side of a worm-like Spider-mite, which lives as a parasite, buried in the skin at the roots of the hairs of animals, and thereby forming a pustule; *m*, the mouth; *cl*, the four pairs of feet; *ab*, the abdomen. — *Originally from Simon*.

Fig. 176 (G). *Epeira trifolium*. Hentz. Natural size. For description of lettering see fig. 129, p. 220.

Fig. 177 (H). *Myzostoma cirrhiferum*. Leuck. Natural size about $\frac{1}{10}$ of an inch long. An external parasite on a kind of starfish (Comatula). A view from below. *m*, the mouth; $m^1$, the proboscis protruded from *m*; $m^2$, the aperture of $m^1$; *ph*, the throat; *ph* to *an*, the whole length of the intestine; *v*, $v^1$, $v^2$, the branching prolongations from the intestine; *n*, the concentrated nervous mass, which gives off branches toward the head and tail, and to the five pairs of hooked feet (*cl*); $ov^2$, $ov^3$, $ov^4$, the right and left V-shaped male reproductive organs; *ov*, $ov^1$, the apertures of the last; *pe*, $pe^2$, the right and left oviducts through which the eggs (*w*, *w*) pass when laid; $pe^1$, the common aperture of *pe*, $pe^2$; *cl*, the five pairs of short, jointed, hook-tipped legs; *sk*, $sk^1$, $sk^2$, the four pairs of suckers, or sucker-like organs of adhesion.

Fig. 178 (I). *Tardigrada*, sp.? Duj. 160 diam. A profile view of a Tardigrade which has four pairs of short, clawed legs (*cl*); *m*, the mouth; *th*, the throat; *j*, the gizzard or internal jaws; *st*, the intestine; *an*, posterior end of *st*. — *From Dujardin*.

Fig. 179 (J). *Lydella*, sp.? Duj. 600 diam. A Tardigrade which has four pairs of long and distinctly jointed legs (*cl*); *th*, the throat; *j*, the gizzard-like internal jaws. — *From Dujardin*.

CERTAIN GROUPS OF ARTICULATA. 273

not only on account of its shape, but also because it lives in the intestines of a serpent, the Boa-constrictor ; but, by the researches of Van Beneden, it was ascertained that in the youngest stage of growth, just at the period of hatching from the egg, it has the form represented here (B). This at once convinced its discoverer that the Linguatula is not a worm; but yet it does not appear to have proved the relationship of the animal to any particular group; for although Van Beneden says that the young have a close affinity with those of *Anchorella*, an animal well known to be a Shrimp, he is not positive as to this relationship; and, on the other hand, there are naturalists who think that the young Linguatula exhibits fully as much the characters

Fig. 180 (K). *Prostomum lineare*. Œst. For the lettering and description see p. 36, fig. 19.

Fig. 181 (L). *Albertia vermiculus*. Duj. 100 diam. A view from below. $m$, the mouth ; $j$, the internal jaws; $c$, glands attached to the sides of the throat, probably for the purpose of applying saliva ; $p$, the throat; $g$, lateral pouches of the stomach; $cv$, the vessels of the aquiferous circulatory system; $e$, $e^1$, the eggs; in $e^1$ the young is far advanced in development; $t$, the tail. — *From Dujardin*.

Fig. 182 (M). *Dendrocœlum lacteum*. Œst.? 3 diam. For the description of the lettering see p. 92, fig. 47.

Fig. 183 (N). *Conochilus volvox*. Ehr. From fresh water. Natural size 0.260 millimetre long. A worm-shaped Rotifer. $lb$, the cilia-bearing lobes of the head; $an$, the feelers; $t$, $t^1$, the tail; $m$, the mouth; $m^1$, cavity of the mouth; $j$, the jaws; $st$, $st^1$, $st^2$, the three apartments or subdivisions of the stomach; $o$, the pair of eyes; $eg$, an egg; $ms$, longitudinal muscles used to retract the body into the gelatinous sheath in which the animal lives. — *From Cohn*.

Fig. 184 (O). *Squamella oblonga*. Ehr. 200 diam. From fresh water. A view from below of a Rotifer which has a distinct shell or carapace ($s$, $s^1$, $s^2$). $s$, the anterior transverse edge of the carapace ; $s^1$, the anterior, and $s^2$, the posterior corners of the carapace; $s^3$, the border of the oval, flat area which occupies the lower face of the carapace; $lb$, the cilia-bearing lobes of the head; $t$, the fork of the tail ($t^1$) ; $m$, the mouth; $j$, the jaws; $j^1$, muscles which move $j$, $st$, the stomach; $cv$, the contractile vesicle, or heart of the aquiferous circulatory system; $cv^1$, $cv^2$, the right, and $cv^3$, $cv^4$, the left aquiferous circulatory vessels; $eg$, $eg^1$, and $eg^2$, two largely developed young. — *Original*.

Fig. 185 (P). *Cyclops quadricornis*. Müll. 50 diam. For the lettering and description see p. 220, fig. 128.

18

of the Mites, (a group of Spiders,) as it does those of Shrimps. It is a curious fact, though, that the male (C) of one of that group of Shrimps, the Lerneans, which live as parasites upon the bodies of various aquatic animals, has such a strong resemblance to the Mites, whilst the female of this same species (*Anchorella uncinata*) has an extraordinary worm-like shape and proportions, and yet, in its younger stages of growth, exhibits such characters as incontestably bring it in close proximity to that group of Shrimps which are known as the *Cyclopidæ* (P). As an instance of diversity of opinion upon this point, I would state, also, that Van Beneden thinks that the Mites, instead of being classed with the Spiders, should be placed with the *Lerneans* (C), and that the latter, being removed from the neighborhood of the *Cyclopidæ* (P), should form, in connection with the Tardigrades (E, I, J), a separate class among the Articulates.

Now, in regard to the group of Tardigrades (E, I, J), I would say, that, although there has been an inclination to class them with the Worms, yet the strongest tendency has been to place them in the neighborhood of the Mites; and this would seem to be further warranted since the discovery of this animal, (J) *Lydella*, with its eight long, distinctly jointed legs, and hooked claws. Although this other animal, (I) *Tardigrada*, has the elongate form of the Worms, yet its resemblance to one of the similarly worm-like Mites, as represented here, (F,) a *Demodex*, is equally striking; and moreover the internal organization partakes much more strongly of the character of the latter. It would seem, therefore, that, although the worm-shaped *Linguatula* (A) appears to be linked with the Lerneans (C), Tardigrades (E, I, J), and Mites (F), which Van Beneden would unite in one *class*, yet it is equally clear that the same animal may be traced through the Mites (F) to the Spiders (G), or through the Mites (F) and Tardigrades (E, I, J) to the Spiders (G), or, finally, through the Lerneans (C) to the Cyclopidæ (P), and from the latter to the Lobsters and Crabs.

There is another worm-shaped animal, (L) *Albertia*, which lives as a parasite in the intestines of Snails and Earth-worms.

CERTAIN GROUPS OF ARTICULATA.     275

Not only from its form and habit is it entitled to be called a parasitic worm, but also because its internal organization is in a large degree like that of the true intestinal worms. On this account the group to which it belongs, the *Rotifera* (L, N, O), is most decidedly affirmed by some naturalists to be allied in every respect to the Worms; but there are others who are equally tenacious in their belief that they should be classed with the Shrimps (Cyclops, &c. P), Lobsters, and Crabs. This they urge upon the ground, that, although some of the Rotifers are worm-like in form, others, and the majority of them, have a more or less hardened shell, (O, $s$, $s^1$, $s^2$,) which in a large number of cases is nearly as broad as long, like that of Lobsters and Crabs; and that they produce a peculiar brood of eggs, known as the "*winter-eggs*," which they carry around with them attached to the outside of the shell, in the same way as the Cyclopidæ (P).

I come now to an animal, *Myzostoma*, represented by this figure, (H,) which of all creatures known presents the most puzzling combination of characters, not only externally, but also in its internal organization. It lives as a parasite upon a species of Star-fish. If we consider its general form, in connection with its five pairs of jointed, foot-like appendages, (*cl*,) which are tipped by hooked claws, or bristles, its stomach, (*ph* to *an*,) to which are attached numerous, irregularly branching appendages, ($v, v^1, v^2$,) and the union of the male and female organs of reproduction (*ov* to $ov^4$, and *pe*, $pe^1$, $p^2$, *w*, *w*) in the same individual, we are reminded of the *Tardigrada* (I) ; or if we add to these characters the highly concentrated nervous system (*n*), and the remarkable similarity of its young to those of the Mites, we cannot fail to see its close resemblance to the latter. But if again we take into account its protrusile proboscis, ($m^1$,) in connection with the branching stomach, (*ph* to *an* and $v, v^1, v^2$,) its peculiar organs of reproduction, which are both male ($ov, ov^1, ov^2, ov^3, ov^4$) and female (*pe*, $pe^1$, $pe^2$, *w*, *w*), and its four pairs of adhering suckers (*sk*, $sk^1$, $sk^2$) ranged along the under side of the body, we are led to refer it to the same group as the *Planariæ* (M), or some of their vermine congeners, *e. g*. Prostomum (K), or the parasitic Distomas, Tetrastomas, Octobothriums, &c. On the whole, it would seem that this strange anomaly contains many more of the worm-like features than of others, and if we add, to the last-mentioned group of characters, the short feet, (*cl*,) with their protrusile bristles, we may assume that we have before us a peculiar kind of worm, combining in itself the organization of the Intestinal worms and the true *Annulata*, or Aquatic worms, and therefore standing as an intermediate form, or group, between the former and the latter.

Such are, in brief, the links which unite these apparent anomalies of nature. Several of them have occupied the attention of zoölogists for many years; and notwithstanding their anatomical characters have been reviewed again and again, their definite position among the various groups of Articulata remains undetermined to the present day. Those who are satisfied that animals are related to each other through inosculating, or *interlocked groups* are content with their present understanding of the general relations of these forms; but that class of naturalists who think they see in every kind of animal a distinct, circumscribed idea, will always be straining their nerves to the utmost to fit every stray piece to its exact mate in the gigantic puzzle.

As we rise from the consideration of general relations to the more specified and circumscribed, *i. e.*, from those which run through the whole animal kingdom to those which more especially characterize the great groups or divisions, and from these pass on to the characteristics of the classes, (or subdivisions of the great groups,) we find the features of the minor groups have more of the nature of what are commonly called *family traits*, or such as would be usually recognized as the physiognomy of near blood-relationship.

But now we have, as it were, only just begun to get an insight of the still closer and more numerous affinities which are common to the various *species* of animals. Among the Song-sparrows, for instance, who can say how many characters they have in common? Beside the more general characters which run through all animals, and the characters of the Vertebrate type, and the Bird features of their external conformation, there are the characters which unite them in the little group called the *Song-sparrow group*. But now, in regard to the latter case, how many different kinds of Song-sparrows are there? Some say four or five, and others say seven; and even the same author is in doubt to a certain extent. If the *seven* so-called species can be reduced to *four*, then there are *three* of them which must have arisen from a *common parentage* with certain ones of the *four;* but if the *seven* are really distinct, then each had a separate parent. Now these closely related specific diversities are more or less such as may be exhibited even in the same nest; and it is on this account, and with the knowledge of these facts, that a large number of naturalists have argued that the diversities among the different species have originated from a common parentage; and since, too, they have observed that the group of Song-sparrows are more or less intimately related to the Chipping-sparrows or Buntings, and in fact cannot be separated from them by any definite line, they refer the two groups to some primeval parentage; — and so they go on linking, by this method of comparison, one group of birds with another, in various degrees of relationship, until the whole class of Birds is referred to *one typical feathered creature.*

But now comes the question as to the relation of the *typical bird* to the other classes, the Mammals on the one hand, and the Reptiles and Fishes on the other. I have already spoken of the transition from the Fishes to the Reptiles in the instance of the Lepidosiren (p. 263); and now I have to present in regard to the Birds another just as remarkable case of relationship. There has recently been discovered in the lithographic stone region of Germany a vertebrate, fossil animal, which combines such a set of characters in its skeleton, and the impressions of its feathers, as to induce in those who have seen it the same diversity of opinion in regard to its relations as was done in the case of the Lepidosiren. On the one hand, a certain German naturalist pronounced it a *bird;* but another characterized it as a *feathered reptile.* Richard Owen, having obtained possession of it for the British Museum, has just issued a memoir in which he describes and illustrates this interesting fossil under the name *Archeopteryx,* with the fullest details; and, in summing up its singularly diversified features *pro et contra,* he designates it as a *bird with strong reptilian characters.*

Not all Vertebrates that have wings are Birds. The Bats are a numerous group of warm-blooded Quadrupeds which fly as rapidly as most birds. In past geological ages there were certain Reptiles, like this one here (fig. 186) known as the *Ptero-*

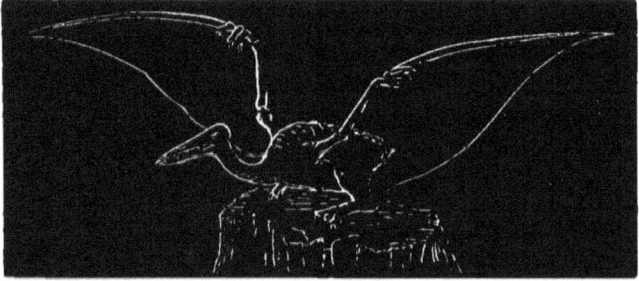

Fig. 186.

*dactyls,* which evidently had a power of flight equal to that of the Bats.* Some of the Pterodactyls were gigantic, and, as Richard Owen has estimated, " had a probable expanse of wing of from eighteen to twenty feet."

Now did time allow I could go on and show how and where occurs the passage from the Birds to the Mammals, or from the Reptiles to the Mammals; but I will only refer to one of these

* It has been argued by some that as the Pterodactyls have scarcely any median ridge to the sternum, they could not possess such a basis of attachment for the pectoral muscles as would be sufficient to sustain them in flight; and that therefore the wing-like fore-arms were used merely as paddles to swim with. Let any one inspect the sternum of the Bat, and he will there find, in its very low median keel, just as good an argument why that animal should not fly, — although we know that it does, — as in the case of the Pterodactyl.

Fig. 186. *Pterodactylus.* A flying Lizard which lived in the middle or Mesozoic age. A copy of a figure which was drawn from a model constructed under the supervision of, and after a restoration by, *Richard Owen,* and set up in the gardens of the Crystal Palace at Sydenham, England.

instances in a brief manner, and without going into many details. The *Duck-mole*, (p. 53, fig. 26,) or Ornithorhynchus, is the lowest of all living Mammalian quadrupeds, and has such a peculiar combination of characters in its organization as to have induced Lamarck to pronounce it as neither a Quadruped nor a Bird. Not only are its jaws toothless, but they are covered by a horny sheath shaped like a duck's bill; its young are born in a *soft shell*, in the same way as are the young of birds and reptiles, and it has spurs on its hind legs like a fowl; and, finally, in correspondence with these characters, its internal organization, in some points, is almost identical with that of birds.

What, then, shall we say of all these degrees of relationship; shall we admit that the child is like the parent because the parent gave it birth, and transmitted its features to it as an *hereditary link*, or must we look upon the parent as the mere impotent covering by which the young is protected whilst it develops certain foreordained characters, the most remarkable of which is the resemblance to the animate enclosure from which it finally emerges? When we admit that the several members of one family, which had one common parent, have *individual differences* by which they can be distinguished, the one from the other, we begin to acknowledge that the same parent does not invariably project the same line of influences upon each of its offspring, at each successive birth. Let now the influence, in any one line, be projected by the offspring upon *its* young, and each of the several offspring of the latter project the influence which it has inherited, and each one of the resultants of these, at the same time, transmit another set of influences, and so on from parent to child, and child to grandchild, and from grandchild onward *ad infinitum*, and in the course of time the offspring of the long past progenitor would hardly be recognized as of the same race.

Thus I think it is, that, by the multiplication and intensifying of individual differences, and the projection of these upon the branching lines of the courses of development from a lower to a higher life, the diverse and successively more elevated types of each grand division have originated upon this globe.

# PART THIRD.

THE MODE OF DEVELOPMENT OF ANIMALS CORRESPONDS WITH THE TYPE OF THE GRAND DIVISION TO WHICH EACH ONE SEVERALLY BELONGS.

# PART THIRD.

THE MODE OF DEVELOPMENT OF ANIMALS CORRESPONDS WITH THE TYPE OF THE GRAND DIVISION TO WHICH EACH ONE SEVERALLY BELONGS.

## CHAPTER XVII.

THE SEGMENTATION OF THE EGG. — THE EMBRYOLOGY OF PROTOZOA, ZOÖPHYTA, MOLLUSCA, ARTICULATA, AND VERTEBRATA.

As a concluding lecture of this course, I propose to show that the *mode of development of an animal* corresponds with the *type of the division* to which it belongs.

Of course it would not be possible in so brief a period to present anything more than a sketch or outline of the *processes* which are involved in the origination of the several parts of the animal organism; nor is it really necessary to go into the minuter details in order to reach the characteristic features which enable one to distinguish the several types from one another.

I have already told you that the relation of the parts of an organism constitutes its *type*. You have seen the relation to a spiral in the Protozoa; the repetition of parts across the body in Zoöphyta; the uniformity of organization in the Mollusca; the repetition of parts from point to point along the axis of the body in Articulata; and the body divided into an upper and lower region by a vertebral column in the Vertebrata. These are characters which are not hidden in the minute structure of each particular organ, but lie on the surface as it were; and this is none the less true of the *embryos* of animals.

I need but refer to the progressive steps in the formation of the egg, to recall to your minds what I have said in regard to the principle of *bipolarity* with which the eggs of all animals begin life. This is a principle which is evinced to the latest period of the life of the *adult*, in the relation of *back* to *front*, or, as it is more frequently called, *above and below*.

There is one *peculiar* feature in the development of the *egg-stage*, which I have not as yet said anything about. I refer to the *self-division* of the egg. It is a well-demonstrated fact, that, just before the organization proper begins to form, the egg of all animals divides itself into a greater or less number of minute spheres; and in doing this it adopts a process which is most remarkable as an instance of *studied regularity* in such a simple and apparently formless body.

There is an almost infinite number of varieties or modifications of this process; but I cannot at present do more than display the essential features. I will take for my illustration the egg of *Laomedea*, which I have already described on a previous occasion (p. 33, fig. 14). At first each egg divides into two more or less distinct halves (fig. 187, A); and at the moment this begins, the germinal vesicle bursts, and its albuminous contents are diffused and mixed with the surrounding material. As a second step, each half again divides; and thus is produced a quadruple body (B), instead of the simple one of a few minutes before. Then each of the four divides again; by this increasing the number of the segments to eight (C). After this it is difficult to follow the process of segmentation, on account of the overlapping of the little spheres, and their opacity; but still we can readily observe that they are

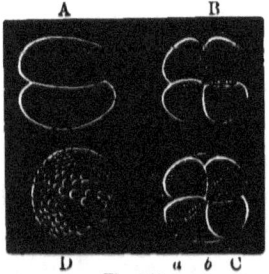

Fig. 187.

Fig. 187. The segmentation of the egg of *Laomedea amphora*. Ag. 125 diam. A, the first division into two masses; B, the second division into four segment spheres; C, the third division in eight masses, of which four (*a*) are nearest the eye, and four (*b*) alternating with *a*, lie in the distance. D, the so-called mulberry-mass, which results from the continued repetition of the self-division of the segment spheres. — *Original*.

THE DEVELOPMENT OF PROTOZOA. 285

increasing in number and consequently diminishing in size (D) as time passes on. By and by, however, having become a multitude in number, and exceedingly diminutive in proportions, they cease to divide.

After the egg is thus converted into what is called the *mulberry state*, the process of developing the organs begins, and is carried out in a manner which varies according to the type to which each several animal belongs.

§ PROTOZOA.

It is so short a time since it was discovered that the Protozoa reproduce themselves through the mediation of the egg-phase, and the subject is such a difficult one to follow out, on account of certain peculiarities in the habits of many of the members of this group, that we have only been able to get a glimpse of the mode of development which is peculiar to this class. In fact, it will not be possible, on this occasion, to do more than indicate two or three phases, at most, in the development of the young from their earliest to their latest period of growth. As far as they go, however, they clearly demonstrate that the process of development is totally unlike that of any of the other grand divisions, and in strict conformity with the typical idea upon which the Protozoan organization is based.

The Protozoa consist of two clearly marked groups. The one begins with Amœba, (p. 9, fig. 1,) Difflugia, (p. 11, fig. 2,) Cornuspira, (p. 13, fig. 3,) and Rotalia, (p. 14, fig. 4,) and rising through the successively higher forms, such as Sponges, (p. 41, fig. 21,) Actinophrys, (p. 44, fig. 22,) Lithocampe, (p. 49, fig. 23,) and Zoöteira, (p. 50, fig. 24,) finally culminates in Podophrya, (p. 51, fig. 25). The members of this group do not possess vibratile cilia as a means of locomotion in their adult state, but are dependent altogether upon certain organs of adhesion for this purpose, and for seizing their food. Certain forms have been classed together under the name *Suctoria*, on account of their habit of sucking the juices of their prey through their feelers. Such a one is Podophrya (fig. 25); but as I conceive that it stands, with many others of a similar kind, in direct relation with those which I have just mentioned, not only by its method of feeding, but also by its internal organization, it would seem to be advisable to apply the name *Suctoria* to the whole group, from Amœba to Podophrya, merely for the sake of having some single term by which we may distinguish it from the other group, the *Ciliata*.

The *Ciliata* is a group which comprises all those usually free forms that are endowed with one or more vibratile cilia, and includes, in an ascending series, Euglena, (p. 144, fig. 86,) Chlamidomonas, (p. 145, fig. 87,) Heteromastix, (p. 146, fig. 88,) Ceratium, (p. 148, fig. 89,) Pleuronema, (p. 148, fig. 90,) Paramecium, (p. 163, fig. 96,) Stentor, (p. 62, fig. 30,) and attains its highest point in Epistylis (p. 161, figs. 94, 95).

*Suctoria.* As a representative of the mode of development of the first group, I shall bring forward one of its highest members, *Podophrya*, (p. 51, fig. 25,) not only because it is best known, but also for the reason that it presents the greatest range of modifications in the process of development. All that has been learned of its earliest stages is that the embryo appears as a clear globular body within the parent. Next it is observed to have a wreath of vibratile cilia about an oval, or broad, cylindrical body. In this condition it is born, and, when fully expanded, (B,) its wreath of cilia is seen to be arranged in an oblique circle about one end, and two contractile vesicles become evident, one on each side of the body. In a short time the vibratile cilia disappear, and four groups of globe-tipped feelers (A) take their place. When the young has developed a stem, which it does in a very few hours, it has all that is requisite to constitute it a fully organized Podophrya, although, before it arrives at a full adult state, the size of its body, as well as the number of its feelers, increases considerably.

*Ciliata.* There is no one of the members of this group whose embryology has been so closely and frequently studied as that of *Paramecium* (p. 163, fig. 96). It is a curious fact in its developmental history, that it passes through a course which appears to be the reverse of that of Podophrya (fig. 25). After undergoing a peculiar kind of segmentation, the egg is changed to a simple globule, which, for want of a distinguishing feature, might remain unnoticed, were it not associated with numerous others of the same kind in the reproductive organ. In process of time this globule becomes endowed with a clear, hollow space, which, by its regularly alternating contraction and expansion, may be recognized as a contractile vesicle. This is the earliest condition in which the embryo can be distinguished, as such, in an isolated state. Presently there is added to this a covering of cilia, of two kinds; the one, however, usually preceding the other, although with some irregularity. The one kind are simple globe-tipped threads (fig. 188, $f, f$) which are very few in number, and do not vibrate; the other kind are like ordinary vibrating cilia, and are numerously distributed over the whole body. At the period of birth sometimes only the globe-tipped threads ($f, f$) are present, and the vibrating cilia appear afterwards. The young Paramecium represented here (fig. 188) possesses both kinds of cilia, and, moreover, exhibits a trace of the characteristic oblique mouth ($m$) at one side of its otherwise symmetrical body. To complete its growth it needs to drop the globe-tipped cilia, develop another contractile vesicle near the posterior end, perfect the organization of its, as yet, incipient digestive system, and increase its size and proportions to a considerable extent. It has already developed far enough to enable one to recognize in its features the Protozoan type; but it does not possess such definite characters as would distinguish it from those other kinds of infusorians which are closely allied to the one from which it originated. It is not always an easy matter to distinguish one kind of adult Paramecium from another, and much less readily can it be done with the younger. The only means that we

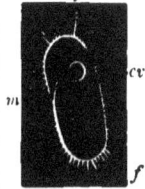

Fig. 188.

Fig. 188. The embryo of *Paramecium Bursaria*. Focke. 350 diam. $m$, the first trace of the mouth; $cv$, the contractile vesicle; $f, f$, the globe-tipped cilia. — *From Claparède.*

possess of knowing to what species an embryo belongs is either to see it at the time of its birth, or watch it until it has grown to an adult condition. The former is a comparatively easy task; but the latter has as an obstacle to such a consummation the element of time, and the difficulty of keeping such minute creatures in view, and alive, for any considerable period.

The young of *Stentor* passes through a series of changes which differ in some respects from those of Paramecium; for instance, it begins as a simple globule, and then develops within it a single, globular, contractile vesicle; but neither in the next phase, nor at any time, are there to be seen such globe-tipped cilia as are so conspicuous on the young Paramecium. At the

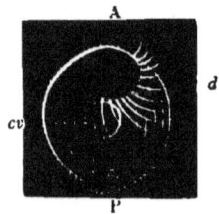

Fig. 189.

time of its birth the embryo Stentor (fig. 189) possesses two sets of vibratile cilia, which are disposed in such a manner as to give it a strong resemblance to some of the Protozoans of a lower rank, *e. g.*, Pleuronema (p. 170, fig. 99); and in fact the shape and proportions of its body are such that one could hardly be persuaded that it had the most distant relationship to its parent, were it not possible to see that it possesses these characters before it is born. Of its two kinds of cilia, the smaller are spread universally over the body in longitudinal lines, and the others, possessing more of the character of lashes, are disposed in a very oblique, arched line, (fig. 189, *d*,) which extends, from near the anterior end of the body, backwards, first toward the left side, and then transversely along the lower face to a point a little to the left of the middle line, where it terminates abruptly. The space embraced by this arch corresponds to the disc of the adult, (fig. 30, *s*,) but as yet it lies nearly in the same plane with the rest of the body, like the homologous region of the most oblique, adult forms of this group. In progress of

Fig. 189. The embryo of *Stentor polymorphus*. Ehr. 350 diam. A view of the inferior side of the body. A, the anterior, and P the posterior end; *d*, the semicircle of vibratile cilia at the edge of the disc; *cv*, the contractile vesicle. — *From Claparède.*

time, however, the disc becomes completely surrounded by the arch of cilia, ($d_1$) and during the process gradually assumes a position similar to that of the adult, *i. e.*, obliquely transverse to the longitudinal axis of the body.

§ ZOÖPHYTA.

Of the four classes of this division, namely, the Polypi, Acalephæ, Ctenophoræ, and Echinodermata, I shall illustrate the development of only the first and the last, which are respectively the lowest and the highest of the group. The mode of growth of the young of the Acalephæ has been sketched already when treating of the process of self-division, and the orientation of the tentacles of Aurelia and its *ephyra* and *scyphostoma*.*

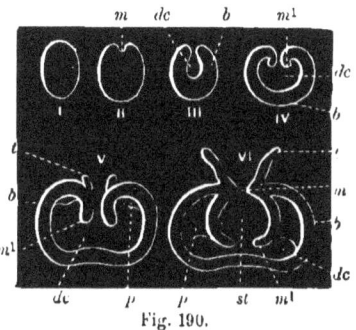

Fig. 190.

*Polypi.* After undergoing the usual process of segmentation, the egg of the Sea-anemone (p. 57, fig. 28) loses its peculiar character, and becomes transformed into a solid oval mass (fig. 190, I). Presently one end of it becomes indented (II, $m$); then in process of time the indentation deepens until it reaches the

Fig. 190. The several stages of development of a *Sea-anemone*, slightly magnified. I, the first stage after the segmentation of the yolk. II, the second stage; $m$, the incipient digestive cavity. III, the third stage; $dc$, the future digestive cavity; $b$, the solid, vase-like body. IV, the stomach proper beginning to form by an inrolling ($m^1$) of the rim of the vase; $dc$, the digestive cavity; $b$, the base. V, the stomach in an advanced state of development; $m^1$, the posterior edge of the stomach; $dc$, the digestive cavity; $t$, the feelers; $p$, the incipient semi-partitions; $b$, the wall of the body. VI, the stomach completed; $m$, the mouth; $m^1$, the posterior opening of the stomach ($st$); $dc$, the digestive cavity; $p$, the semi-partitions; $t$, the feelers; $b$, the wall of the body. — *From Cobbold.*

* See p. 241 for the description of the earliest stages of development, and p. 67 for the completion of the process, up to the adult condition.

19

centre of the body (III, *dc*); next, the body continuing to increase in size, and the indentation enlarging at a corresponding rate, until it becomes a great cavity, the edge of the indentation appears to be folded or rolled inwards (IV, $m^1$); and finally, we find this incurved edge extending (V, $m^1$) half-way to the bottom of the cavity (*dc*) so as to form a sort of passage-way from the latter to the outside, and at the exterior end of this passage-way several little protuberances (*t*) arranged about it in a single row, whilst within the newly formed cavity (*dc*) there may be seen several low ridges (*p*) which divide the space immediately about the passage-way into as many little stalls. At this latter stage of development we may recognize the characters by which the embryo can be identified as one of the Polyps, so that if we were not cognizant of its origin there could be no hesitation as to its typical relationship. For instance, that which I have called the passage-way corresponds to the stomach, and the cavity, (V, *dc*,) which commenced as an indentation, is the digestive cavity, whilst the low ridges (*p*) in the latter indicate the position of the primary set of semi-partitions, and the protuberances (*t*) around the exterior opening, or mouth, of the stomach stand in the place of the first group of tentacles. In the next and last stage (VI) which I shall illustrate, the organs have assumed very nearly the form and proportions of the younger phases, which I have described in former lectures (pp. 59, 177); thus, the tentacles (*t*) are prominent finger-shaped bodies; the stomach (*st*) is narrowed in front, where the mouth (*m*) is, and behind, at the entrance ($m^1$) to the digestive cavity (*dc*); and the semi-partitions are distinct membranes (*p*) which extend nearly the whole length of the body and divide the space about the stomach into totally separate apartments, whose only means of communication with the rest of the body-cavity are through the ends which lie next the posterior part of the embryo.

*Echinodermata.* I shall go no farther with the illustration of the mode of development of the group of Echinoderms than is sufficient to show that the various organs originate only at such points as have the same relations to each other that we find in the adult. The embryos which I am about to describe were hatched in March, from the eggs of *Holothuria tremula*, a Trepang which lives in the seas about the coasts of Norway. Like the scaly Trepang, *Psolus phantapus*, which I alluded to in a previous lecture, (p. 192, fig. 117,) it has rows of sucker-like feet, on the lower side of the body, and great branching feelers projecting from the head; but the upper side of the body is soft, and covered with numerous protuberances. Interiorly the organization of this Holothurian is, in a general way, like that of the flesh-colored, footless Trepang, *Caudina*, (p. 187, fig. 114,) of our coast; and the latter, therefore, may serve as a means of comparison in tracing the relations of the developing organs of the young of the former.

The mode of formation of the parts in the earliest stage of growth, after the egg has passed through the phase of self-division, is very much like that which occurs in the Sea-anemone; *i. e.*, the globular so-called mulberry-mass becomes indented at one end and finally hollowed within, so as to form a sort of vase with a narrow mouth. In the course of a week the body assumes an oval form, and several of the organs about the head begin to take on distinct shapes and outlines. In this figure, (fig. 191,) which represents the upper side of an embryo at the age just mentioned, the mouth (*m*) is in profile at the extreme end of the head, and just behind it three of the five incipient feelers are visible, standing in such a position that one is in the mid-line of the back, and the others are respectively right and left of it. As we shall see presently in an older phase, the remaining two of the five have the same relation to the inferior mid-line as the superior right and left ones have to the upper median line. A short distance behind the upper odd tentacle, and in the same line, there is a little pit (*mc*) which forms the end of a canal within, that extends from it to a hollow ring or channel about

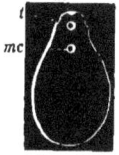
Fig. 191.

Fig. 191. *Holothuria tremula*. Gunn. Magnified. An embryo soon after birth, seen from the back. *m*, the mouth; *t*, the incipient feelers; *mc*, the first trace of the madreporic body.—*From Koren and Danielssen.*

the mouth. The canal corresponds to the madreporic canal which I have described in the Starfish (p. 182, fig. 111, $m^1$) and Caudina (p. 187, fig. 114, $mc$), and the pit is identical in relative position with the madreporic body (fig. 111, $m$, fig. 114, $m$) of the same.

In another week traces of a pair of sucker-like feet (fig. 192, $s$,  $s^1$) become visible on the lower side of the body, and near its posterior end. They have the form of hollows, and are arranged so that one is on the right, and the other on the left of the inferior mid-line, and close to it.

Fig. 192. At this period the feelers ($t$, $t^1$) are quite prominent, and have a club-shaped form. This figure, which is a view of the lower side of the body, displays their arrangement about the mouth ($m$) when they are retracted within their sockets. One ($t^1$) is in the dorsal mid-line, and the others are placed symmetrically right and left, at equal distances apart. Interiorly the organization has a corresponding degree of development; the intestinal canal is a complete digestive organ, with an anterior and posterior opening; the five tentacles of the second set have begun to develop at alternate points with those of the first set; and the madreporic body has become completely covered over by the skin, in fact, totally included, with the madreporic canal, within the general cavity of the body, and stands in the same relation to the other internal organs, as that of Caudina (p. 187, fig. 114, $m$).

If we follow the development of the embryo for a month longer from this time, we shall see the organs assume the positions and proportions represented in this figure (fig. 193), and be enabled thereby to ascertain their homological identity with those of the adult Trepangs. We find here the following organs, namely, the first set of five tentacles ($t$, $t^1$) not much larger than the second, and the members of both branched considerably at and about

Fig. 192. The same as fig. 191, but a little older. A view of the abdominal side. $m$, the mouth; $t$, $t^1$, the feelers; $s$, $s^1$, the sockets of the first pair of sucker-like feet. — *From Koren and Danielssen.*

their free ends; the mouth ($m$) encircled by an aquiferous channel, with which the cavities of the tentacles are in open communication, and from which five longitudinal canals ($aq^2, aq^2, aq^4, aq^5$) extend backwards, just beneath the skin, to the posterior end of the body; the sucker-like feet, ($s^2$,) which have been increased in number to four, protruded from the body through the apertures which lie in pairs close to the inferior mid-line, and their interior in direct communication with the middle longitudinal canal by means of a narrow tube ($s^1$) which passes from each of them to the latter; the madreporic canal ($mc$) in the distance, at the dorsal mid-line, and swollen at the end; the posterior end of the intestine expanded transversely into a broad sac ($g^3$), which eventually becomes the double respiratory organ; five longitudinal muscular bands corresponding in position with the longitudinal canals ($aq^2$, &c.); and finally a few transverse or annular muscular bands ($bd$) lying exterior to the longitudinal ones, and which eventually broaden so as to touch each other, thus forming a continuous contractile sheath immediately beneath the skin.

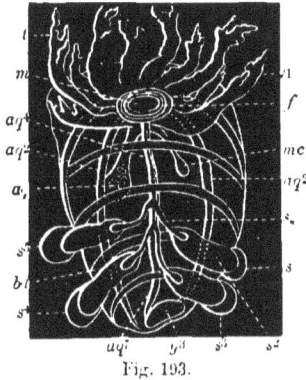

Fig. 193.

What is the process of development from this period onward to the adult state was not ascertained by the investigators, Messrs. Koren and Danielssen; nor is it necessary for our purpose that we should know, since we have followed it far enough

Fig. 193. The same as figs. 191, 192, but a little more than a month older than the last. Natural size about ⅓ of an inch long. A view from below. $m$, the mouth; $f$, the rudiments of the calcareous buccal ring; $t, t^1$, the feelers; $s, s^3, s^4, s^5$, the adhesive disc of the sucker-like reptatory organs; $mc$, the madreporic canal; $aq$, a saccular appendage of the aquiferous ring about the mouth; $aq^2, aq^2, aq^4, aq^7$, the longitudinal aquiferous canals, seen as it were through the longitudinal muscular bands which underlie them; $s^1$, the canals which lead from $aq^4$ to the interior of the sucker-like feet ($s^2$); $bd$, the transverse annular muscular bands; $g^3$, the incipient respiratory branches. — *From Koren and Danielssen.*

to see that it produces an organization in every respect identical in relation with that of the Zoöphytic type, and specially homologous with that of the Holothurians.

§ MOLLUSCA.

One of the most thorough and exhaustive investigations of the embryology of this type is that of Lereboullet upon the fresh-water snail *Lymneus*. This animal is not so very distantly removed in its organization from the common land snail, *Helix* (p. 203, fig. 122) but that the latter may serve to illustrate the general relations of the organs of an adult snail, if we wish to compare the young Lymneus with a mature organization; and this is the more allowable in the present case, because our object is not to trace the development to any particular form, but to show that its general results are in accordance with the type to which the embryo belongs.

The first change that occurs in the globular mulberry-mass, after it has passed completely through its segmenting process, is a flattening and then a hollowing of one side of it until the once solid globule is metamorphosed into a hemispherical cup. Soon, however, the mouth of the cup is narrowed by a mutual approximation of its opposite edges until they form a junction along a straight line (fig. 194, *a*, *m*), excepting at one point, where an oblong aperture (*m*) is left. The aperture is the future mouth, and the cavity to which it leads is the incipient stomach, whilst the line of junction (*a*) of the approximated sides of the cup indicates the trend of the vertical plane which divides the body into symmetrical right and left parts. At the same time, the little globular masses which originally formed the bulk

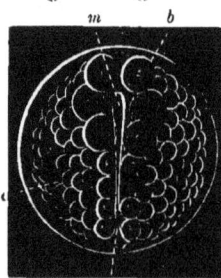
Fig. 194.

Fig. 194. *Lymneus stagnalis*. Lamck. 150 diam. One of the earliest stages of development of an air-breathing *aquatic snail*. *c*. The yolk-mass; *b*, peculiar yolk globules, or segment spheres collected about the incipient mouth (*m*); *m* to *a* the longitudinal axis. — *From Lereboullet*.

of the cup become less conspicuous at the periphery, by an increase in transparency, so that the embryo appears to be enclosed in a hyaline shell. In the course of three days its longitudinal axis becomes more decidedly marked by an elongation of the body in that direction, (fig. 195,) and its bilaterality evinces itself by a symmetrical projection of the right and left sides of the region about the mouth, so as to form a sort of broad tongue ($l$, $l^1$). As the latter is the incipient creeping disc, we are enabled by its position, even at this early period, to determine the upper and lower sides of the body; and as we know the point at which the mouth opens, we can at the same time decide upon the relative situation of the head ($a$) and tail, and right ($c^1$) and left ($c$). So far, then, the embryo, in all its parts, is perfectly symmetrical right and left of the axial plane.

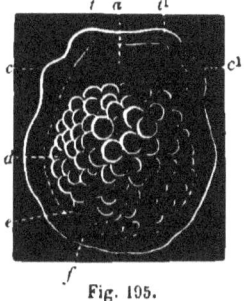
Fig. 195.

In the next stage, which is about twenty-four hours older than the last, the various regions of the body are quite prominent, and new features of an important character have become conspicuous. Premising that the embryo still retains its symmetrical proportions right and left, I will illustrate the relations of its organs by a profile view (fig. 196). The creeping disc ($f$) is quite marked as a distinct swelling which projects below and behind the mouth ($m$). The mantle, which forms a complete envelope to the body in the adult state, has just begun to form, and covers the end of the body, which is exactly opposite to the mouth, with a sort of low cap ($sh$). The digestive cavity appears much

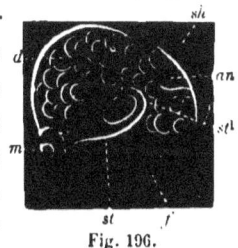
Fig. 196.

Fig. 195. The same as fig. 194, but three days older. 200 diam. A view from the back. $a$, the head; $l$, $l^1$, the right and left corners of $a$; $c$, $c^1$, the right and left sides of the anterior region of the back; $d$, the enclosed yolk mass; $e$, the transparent space about $d$; $f$, the envelope of the body. — *From Lereboullet.*

Fig. 196. The same as figs. 194, 195, but one day older than the last. 100

narrower than in the previous stages, and is divided into two regions, namely, a narrow throat, and the stomach proper (*st*). The latter is as yet closed behind (*st*$^1$), but is evidently undergoing a change preparatory to its connection with the rest of the intestinal tract which is developing isolately at the extreme posterior end of the body. The latter operation commences by a depression (*an*) forming immediately under the cap-like mantle (*sh*), exactly opposite to, and as if in direct continuation of, the stomach (*st*, *st*$^1$). In process of time the depression becomes elongated into a tube, and meeting the posterior end (*st*$^1$) of the stomach unites with it and forms a complete intestinal canal, which projects in a straight line along the axis of the body.

In the course of about three days from this time the first traces of a heart appear, as two slowly pulsating globular sacs placed one before the other, in the midst of the mass of little spherical bodies (*d*) which are so conspicuous in the region of the back during the previous phase of growth. About this time, too, the intestinal canal begins to assume a lateral curvature, and thus becomes the first to depart from the strictly symmetrical relations in which the parts of the organization have hitherto stood, and take on that one-sided character for which the Mollusca possessing a spiral shell are so noted. The unequal development of the other organs soon follows the example of that of the intestine. The cap-like mantle spreads faster on the right side than on the left, and finally covers the whole back; the creeping disc, still retaining its symmetry, elongates and flattens into a foot-like form; the region just above the mouth extends forward and over it beyond the foot, and broadening sideways, assumes the position and office of a head; the region on the back behind the head becomes divided into two portions by the formation of a sort of neck which separates the upper space between the head and tail from the great mass which is directly

diam. A profile view. *m*, the mouth; *st*, the stomach; *st*$^1$, the posterior end of *st*: *an*, the incipient intestine; *f*, the creeping disc; *d*, the remains of the yolk mass, on the back; *sh*, the cap-like mantle. — *From Lereboullet*.

enclosed by the mantle; the two chambers of the heart gradually change their relative position until the receiving chamber (*auricle*) lies obliquely to the right of and behind the distributing chamber (*ventricle*).

All these changes bring the embryo into the condition which is shown in this figure, (fig. 197,) a representation of the young of our most common native Lymneus a short time before it is hatched. Here we find in a profile view the following features, namely, the upper region covered by an asymmetrical, oval mantle (*ml*, *ml*$^1$) which extends at its in-

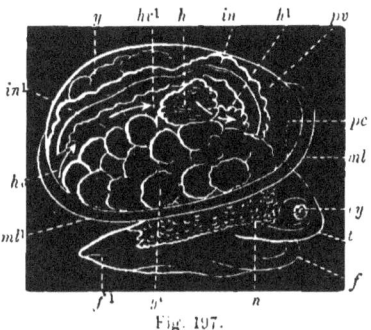

Fig. 197.

ferior edge into a thickened border; the head clearly separated from the creeping disc, (*f*, *f*$^1$,) and furnished with an eye (*ey*) at the base of each of its two lateral, triangular, tactile organs (*t*); the creeping disc shaped like a flat-iron, with the broader end (*f*) in front; the intestine in the form of an irregular slender tube (*in*, *in*$^1$) strongly bent upon itself and arched over the upper surface of the body and to the left of the heart (*h*, *h*$^1$); the circulatory system divided into distinct chambers and vessels, and the blood passing along the back in a current from the main posterior longitudinal channel (*hv*, *hv*$^1$) through a double valve into the auricle, (*h*,) and thence through another valvular opening into the ventricle, (*h*$^1$,) from whence the fluid is thrown into the distributing channels to be carried to various parts of the body; and, finally, the region of the respiratory organ, or *lung*, (*pc*,)

Fig. 197. *Lymneus elodes.* Say. 100 diam. The young of one of our native amphibious snails. A profile view. *t*, tentacle at the corner of the head; *ey*, the eye; *f*, *f*$^1$, the creeping disc; *ml*, *ml*$^1$, the mantle; *pc*, the lung cavity, just behind the front edge of the mantle; *pc*, the region of the lung; *in*, *in*$^1$, the intestine; *h*, the auricle, and *h*$^1$, the ventricle of the heart; *hv*, *hv*$^1$, the posterior blood-vessel; *y*, *y*$^1$, the right and left sides of the back; *n*, the region of the stomach and principal nervous ganglions. — *Original.*

marked out by a clear space at the anterior end of the mantle and near the forward bend of the intestine.

If, now, we compare our young Lymneus with the adult land snail, (p. 203, fig. 122,) we shall see, that, notwithstanding the former lacks, as yet, a distinct nervous and reproductive system, there are other parts of the organization which are unmistakably homologous with those of the latter; and we must conclude, therefore, that at least at this age the embryo conforms to the Molluscan type, whilst we may say, regarding its preceding phases of growth, that there has been nothing seen which did not strictly tend to bring about the result which we have before us. Although it is not necessary for present purposes, I will state that the principal feature which is yet to be developed in order to bring the embryo to the end of its course, is a prolongation of the dorsal region of the body into a long-drawn-out, obliquely conical spiral, closely covered by a similarly shaped shell, within which are included the reproductive organs and the liver.

§ ARTICULATA.

It cannot be repeated too often that the first step which is taken in the development of the young animal, after it has passed through the process of segmentation, is a bipartition of the body into a right and a left side.

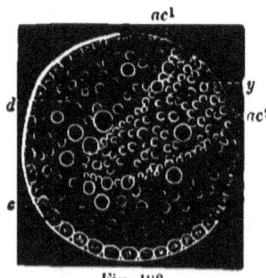

This phenomenon is a very marked feature in the initiatory phase of the embryo of the common black Caddice-fly (*Mystacides nigra*, Latr.) of Europe, a little moth-like insect with very long thread-like feelers and wings slanting like a roof, almost or altogether identical with one which flies into our windows and around the lights in the evenings of early summer.

Fig. 198. *Mystacides nigra*. Latr. 220 diam. The earliest stage of development of the embryo of the black Caddice-fly. *c*, the cellular envelope; $ac^1$, $ac^2$, the edges of *c*; *y*, the yolk-mass not covered by *c*; *d*, clearer portion of the yolk mass. — *From Zaddach*.

The first distinct trace of organization is to be found at the surface, where a well-marked layer of large cells (fig. 198, c) encompasses nearly the whole circuit of the body. At the point where this layer is deficient ($ac^1$ to $ac^2$) the yolk granules ($y$) extend to the surface, whilst that part of the body over which the layer of cells is spread appears lighter and less opaque ($d$). This layer ($c$) might be compared to a ball-cover split open on one side, and then we should have the split trending across the back of the embryo, and one half of the cover representing the head, and the other half the posterior region of the body. You may see from this that the embryo is, as it were, doubled upon itself, and, I will add, in anticipation of what will appear presently, in such a way that its back lies next the centre of the ball, i. e., it is doubled upon itself *backwards*. Which end is the head does not become apparent until the next phase. At the latter period (fig. 199) we find the layer of large cells, the *germinal layer*, and the region which it covers, seemingly narrowed, or contracted upon itself ($ab$) along about two thirds of its length, but at one end ($k$) rather more widely extended than in the previous phase, so that, on the whole,

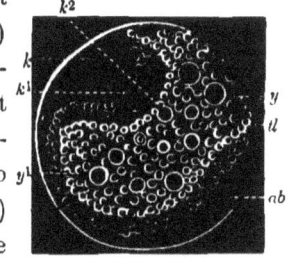

Fig. 199.

when seen in front, it resembles a clumsy sort of Y. We have now, at the broader end of this figure, the first indication of the position of the head, ($k$,) but as yet no intimation anywhere as to whether the embryo is an Articulate or Molluscan, or any one of the five types of animals.

In the next stage, (fig. 200,) however, the true character of the animal, as far as its general typical relations are concerned, begins to develop, under the form of a series of indentations ($i$, $i^1$) which divide the anterior half of the body from point to point along the front or abdominal side, into successive joints

Fig. 199. The same as fig. 198, at a more advanced stage. 220 diam. $k$, the head in profile; $k^1$, the right side of $k$ in the distance; $k^2$, the left side of $k$, next the observer; $ab$, the abdominal region; $tl$, the posterior end of $ab$; $y$, $y^1$, the yolk. — *From Zaddach*.

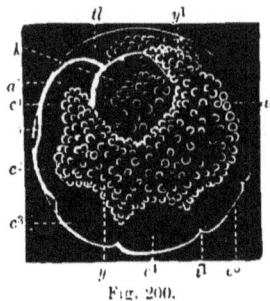

Fig. 200.

$(a^1, c^1, c^2, c^3, c^4, c^5)$. A peculiarity of this phase, not before noticeable, and which is the forerunner of a remarkable change in the relative position of front and back, is evinced by the backward curvature of the head ($k$) so as to sink it inward toward the centre of the yolk mass, ($y_1$) and allow the tail ($tl$) to overlap it. Not yet, though, are we permitted to pronounce upon the special relations of this creature, since the most that its jointed body teaches us is that it is an *Articulate* animal; and, for all we can see, it may be a worm instead of an insect.

But in regard to the latter category, we do not remain long in

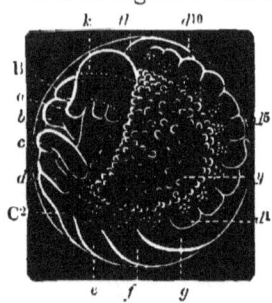

Fig. 201.

doubt; for in a short time three pairs of legs (fig. 201, $e, f, g$) begin to develop from the last three of the joints just mentioned, and in front of them arise four pairs of jointed appendages which appertain to the head, and perform the office of feelers, ($a$,) and upper ($b$) and lower ($c, d$) jaws. The presence of legs determines the character of the embryo as that of an Articulate which is higher in grade than the worms, and their number, namely, three pairs, distinguishes it

Fig. 200. The same as figs. 198, 199, in the third stage. 220 diam. $k$, the head; $ab$, the abdomen; $tl$, the tail, overlapping $k$; $a^1$, the first joint of the body after the head; $c^1$, the second, and $c^2$, the third joint of the same; $c^3$, the first, $c^4$, the second, and $c^5$, the third joint of the thorax, from which the first, second, and third pair of legs develop; $i, i^1$, the transverse furrows between the segments; $y, y^1$, the yolk. — From *Zaddach*.

Fig. 201. The same as figs. 198 to 200, in a further advanced stage. 220 diam. A profile view. $k$, the head; $d^1$, the first, $d^5$, the fifth, and $d^{10}$, the tenth joint of the abdomen; $tl$, the bend of the tail; B, the forehead; $a$, the feelers; $b$, the upper jaws; $c, d$, the under jaws; C², the joint of the body from which $d$ arises; $e, f, g$, the first, second, and third pair of legs; $y$, the remains of the yolk. — From *Zaddach*.

from the spiders, which possess four pairs of legs. In addition to these there are two other characters which, taken in connection with those just mentioned, would, did we not know its origin, vindicate the claim of this embryo to be classed with the *Insects* proper as contradistinguished from the *Spiders*. I mean the isolation of the head (*k*) from the rest of the joints, and the distinct, tenfold jointed division of that part of the body which lies behind the legs, and corresponds to the abdomen ($d^1$ to $d^{10}$). It is here that we see for the first time the beginning of the change in the relative position of front and back which I hinted at just now. This is indicated at two points, and in two separate ways, namely, by the elevation of the head (*k*) from the retroverted position in which it laid at the last stage, (fig. 200.) and by a backward folding of the end (fig. 201, $d^{10}$) of the tail so as to bring it down upon the preceding joints.

The meaning of these movements on the part of the head and tail are fully explained in the next phase, (fig. 202,) where we find the tail (*tl*) completely doubled *under* the body, the head (*k*) standing out as a distinct division, and the back (*y*) projecting above the rest of the body and occupying the region formerly filled by the upturned abdomen; that is to say, the body has completely reversed its relations to the shell in which it is enclosed, by changing from its former retroflexed condition to a conduplicate position. In addition to the increased elongation of the legs (*e, f, g*) and the jaws (*b, c*) and feelers, the complete formation of the rings of the abdomen, ($d^1$, *tl*,) and the appearance of a group of eye-

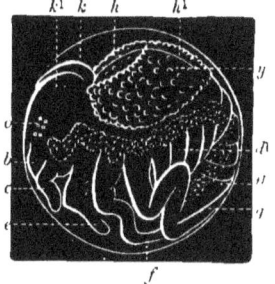

Fig. 202.

Fig. 202. The same as figs. 198 to 201, but more advanced in development. 220 diam. The position is reversed by doubling the tail (*tl*) under the body. *k*, $k^1$, the right and left halves of the head; *o*, the eye; *tl*, the tail; $d^1$, the first joint of the abdomen; *b*, the upper, and *c*, the lower jaws; *e, f, g*, the legs; *y*, the remnant of the yolk, on the back; *h*, $h^1$, a furrow in *y* where the heart develops. — *From Zaddach*.

spots, (*o*,) there is a beginning of the formation of the heart, by a hollowing of a longitudinal furrow (*h*, *h*¹) along the middle of the back, and in the midst of the remains of the yolk mass (*y*).

By one more step we come upon that condition of things in which the proportions of the embryo are nearly the same as those which it has when hatched, and commencing the first of its three phases of self-sustaining life, *i. e.*, its worm state. Although the body has completed its unrolling, the throat (*th*) become a distinctly marked channel, and the furrow (*h*) in which the heart forms is nearly or completely covered over, nothing has been added which renders the embryo any more insect-like than in the last stage; but one or two characters have become developed by which its relations to the Caddice-worms may be inferred.

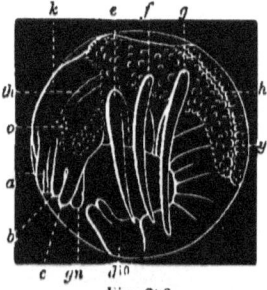
Fig. 203.

The principal of these is exemplified in the forked tail, (fig. 203, $d^{10}$,) upon the prongs of which are eventually developed a hook or claw by means of which the larva clings to the inside of its tubular, aquatic tenement. The other feature is set forth in the very weak, unjaw-like jaws, (*b*, *c*,) which, even in the adult state of the Caddice-fly are scarcely more than feeble organs of prehension, and are called by that name because they occupy the same position as the genuine masticatory organs of other insects, rather than on account of the office which they perform.

When hatched — which happens in the water — the larva constructs about itself a cylindrical tube of bits of sticks, which it glues together with a sort of glairy substance similar to that which silk-worms produce to construct their cocoons. After residing for a time in their portable domiciles, and becoming fully

---

Fig. 203. The same as figs. 198 to 202, just before the period of hatching. 220 diam. *k*, the head; *o*, the eye; *a*, the feeler; *b*, the upper jaw; *c*, the lower jaw; *gn*, the lower lip; *e*, *f*, *g*, the legs; $d^{10}$, the last joint of the abdomen; *th*, the throat; *y*, the remnant of the yolk; *h*, the furrow in which the heart is developed. — *From Zaddach.*

developed as larvæ, they close up the aperture of the tube, and, changing to a chrysalis, lie dormant until the time for their appearance in a perfect state, when the chrysalis-shell is burst open and the perfect caddice-fly takes to its wings.

### § Vertebrata.

Next to those of the common fowl, none of the eggs of Vertebrata are so accessible, and to be had in such large numbers, as those of Turtles and Tortoises; and, insomuch as they do not require, for hatching, more than the natural heat of the ground in which they are buried when laid, they are, of all eggs, by far the most easily preserved in a healthy state during the time of incubation. All that is required to obtain them is to collect a number of turtles in early spring, before May, and keep them enclosed in some shady spot where they can have easy access to water and soft earth, and feed them well with fresh herbage, such as plantain-leaves, lettuce, beet-leaves, &c., &c., and in the course of time, usually in May and June, they may be caught, at early dawn, digging holes in the earth with their hind legs, and depositing therein their brood of eggs, and then covering them up.

As the eggs are required for study they should be taken out one at a time, — carefully, so as not to disturb the others in the least, — and, whilst held in the same position as when in the nest, the shell removed at the upper side so far down as to expose the whole yolk. The eggs of turtles and tortoises have an envelope of albumen, the so-called *white*, very much like that of birds; and in order to keep it, as well as the yolk, in shape, and moist, during prolonged observation, it is necessary to sink the egg in some kind of fluid. For a short period water will suffice; but, as it eventually produces an injurious effect, some denser material, such as thin syrup, which does not react so rapidly, had better be used. The fluid which is left after beef-blood has coagulated is perhaps as good as anything that can be had.

The first trace of an organization presents itself in the form of a round, light-colored disc which lies close to the surface of the yolk. Upon close examination it will be found that this disc

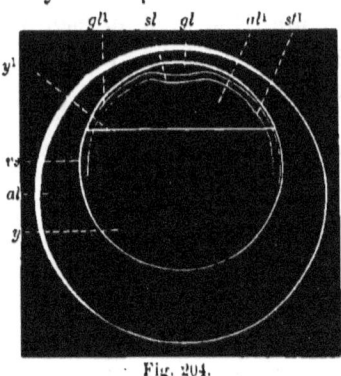

Fig. 204.

is but the thicker portion (fig. 204, *gl*) of a layer ($gl^1$) of cells which has been formed all over the periphery of the yolk (*y*). This layer is called the *germinal layer*, and the thicker, disc-shaped portion of it the *embryonic disc*. Owing to the influx of a portion of the liquified albumen, the region ($al^1$) immediately beneath the disc is very transparent at this time; and, by throwing a strong ray of light through it, we may get such a profile view of the various parts as are represented in this figure (fig. 204).

One of the chief elements in the formation of the organs of the young turtle is the development of what I have called the *subsidiary layer*.* This is a stratum ($sl^1$) of loose yolk granules which originates at the surface of the yolk, and extends all around it. I have called it the *subsidiary layer*, because it is a sort of intermediary between the unappropriated yolk and the definitely fixed parts of the growing organization; in fact, as I

Fig. 204. *Chelydra serpentina.* Schweig. 2 diam. A diagramic, profile representation of the incipient germinal disc of the common *snapping turtle*, as it lies within the egg. *al*, the white of the egg; *y*, the yolk; $y^1$, the surface of the yolk; *vs*, the periphery of the yolk; $al^1$, the clear, albuminous region beneath the disc; *gl*, the disc; $gl^1$, the germinal layer; *sl*, $sl^1$, the subsidiary layer. — *Original*.

* See my observations to this effect in Agassiz's " Contributions to the Natural History of the United States," vol. II. p. 536. I would add, in reference to the " European naturalists" mentioned in the " Note on Scientific Property," p. 38, see Kölliker, "*Entwickelungsgeschichte, &c., Akademic Vorträge*," 1861, p. 20; Gegenbaur, " *Bau & Entwickl. Wirbelthiereier*," *Archiv. fur Anat. und Physiol.*, 1861, p. 497; and also Leuckart, " *Berichte*" in *Wiegman's* (*Troschel's*) *Archiv. für Naturgeschichte*, 1864–5. From American scientific men I have received abundant recognition of my claim, both by correspondence and in public lectures; to say nothing of their free and outspoken sympathy in personal intercourse.

shall show presently step by step, every organ takes its origin from it by a separation of a portion of the thickness of its upper side. I might say, therefore, in the strictest sense, that the *germinal layer* ($gl, gl^1$) is the first to split off from the subsidiary layer. We have, then, in this phase, the following features, namely, the yolk sac ($vs$) partially occupied by liquified albumen ($al^1$) which has infiltrated through its tissue; the whole mass of the yolk (the egg proper) removed from the centre of the white ($al$) and buoyed up near its upper side, taking the place of the absorbed albumen; the *germinal disc* ($gl$) and its peripheric extension, the *germinal layer* ($gl^1$); and finally, the *subsidiary layer* ($sl, sl^1$) from which the germinal layer ($gl, gl^1$) split off. As yet there is no indication of right and left, nor of head and tail, in the germinal disc. These two features develop almost simultaneously. The latter, however, is the first to appear, and may be recognized as a thickening (fig. 205, $hd$) of one edge of the disc, and a sinking of the germinal layer at that point and also at the opposite edge, ($am$,) so that two shallow furrows are formed. The former becomes evident before the latter has completely developed itself, and may be recognized as a faint, straight, shallow furrow which passes through the middle and at the upper surface of the germinal disc ($gl$). This is called the *primitive stripe*, and corresponds to the axis of the body.

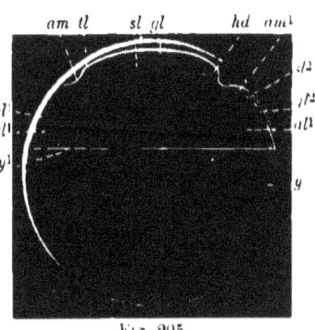

Fig. 205.

In the next phase (figs. 206, 207) the characteristic feature of Vertebrata makes its first appearance; the primary element — the so-called *chorda dorsalis* — of the vertebral column is formed.

Fig. 205. The same as fig. 204, with the shell and white taken off, but further developed. 2 diam. $y$, the yolk; $y^1$, surface of $y$; $al^1$, clear, albuminous space; $gl$, the germinal disc; $gl^1$, $gl^2$, the extension of $gl$ beyond the body; $hd$, the head; $tl$, the tail; $sl$, $sl^1$, $sl^2$, the subsidiary layer; $am$, the furrow around $gl$; $am^1$, the peripheric fold of $am$. — *Original*.

20

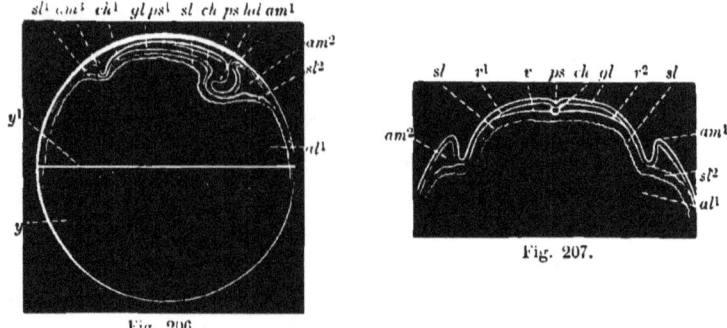

Fig. 207.

In this profile figure (fig. 206) it is the thin strip ($ch$, $ch^1$) which is split away from the subsidiary layer, ($sl$,) except at its anterior ($ch$) and posterior ($ch^1$) ends, and in the transverse section (fig. 207) it appears as the small circle ($ch$) which immediately underlies the primitive stripe, ($ps$,) and is flanked right and left by a thick layer, ($v$,) — likewise split off from the subsidiary layer, — which thins out to a sharp edge ($v^1$, $v^2$) near the margin of the disc. With these additions and the further development of the parts already begun we have the following characters, namely, the infiltrating albumen ($al^1$) occupying the whole upper half of the egg; the germinal layer, not only depressed so as to form a furrow all around the germinal disc, but its fold bent inward (at $am^1$, $am^3$) and elevated so as to partially overlap the slightly incurved and sunken head ($hd$) and tail and sides, as it were, enclosing the embryo in a sort of socket, — the beginning of a development of the *amniotic sac;* the head ($hd$) considerably increased in thickness, or depth; the primitive stripe ($ps$, $ps^1$)

Fig. 206. The same as figs. 204, 205, still more advanced. 2 diam. $y$, the yolk: $y^1$, surface of $y$: $al^1$, the clear, albuminous space; $gl$, the germinal disc; $hd$, the head; $ps$, $ps^1$, the primitive stripe; $ch$, $ch^1$, the *chorda dorsalis*; $am^1$, $am^3$, the peripheric fold of the amniotic sac; $am^2$, the space beneath $am^1$, and closed below by $sl^2$; $sl$, $sl^1$, $sl^2$, the subsidiary layer. — *Original.*

Fig. 207. A transverse section of the body, and the layers immediately about it, of fig. 206. $al^1$, the clear, albuminous space; $ps$, the primitive stripe; $gl$, the germinal layer of the disc; $ch$, the *chorda dorsalis*; $v$, the vertebral layer; $v^1$, $v^2$, the edges of $v$; $am^1$, the peripheric fold of the amniotic sac; $am^2$, the space beneath the fold of $am^1$; $sl$, $sl$, $sl^2$, the subsidiary layer. — *Original.*

furrowed deeper into the head and thinning out, *i. e.*, growing shallower toward the mid-length of the body; the *chorda dorsalis* (*ch*, *ch*¹) lying between the right and left halves of the body, and just beneath the primitive stripe (*ps*); the *vertebral layer*, (*v*,) upon which the back-bone is eventually developed, completely split off from the upper surface of the subsidiary layer (*sl*) and lying in equal parts right and left of the *chorda dorsalis* (*ch*); the subsidiary layer (*sl*) lying close to the under face of the disc, and slightly depressed by the downward curvature of the head and tail, but separated from the rising fold of the germinal layer, and extending directly (at $sl^1$, $sl^2$) to the periphery of the yolk.

In the next phase (fig. 208) it is clear that the embryo has advanced considerably both in regard to the development of those organs already initiated, and by the institution of new features. The body is nearly enveloped by the uprising and contracting edge ($am^1$, $am^3$) of the *amniotic sac*. The head (*hd*) and tail (*tl*) are deeply sunken toward the centre of the yolk mass, (*y*,) and so strongly incurved toward each other as to form a cavity between them, which opens on the lower side into the great area occupied by the albumen, ($al^1$,) and on the upper is bounded by the inferior surface of that portion of the subsidiary layer (*sl*) which lies between the two ends of the body. The furrow of the *primitive stripe* extends more than two thirds of the length of the body, and

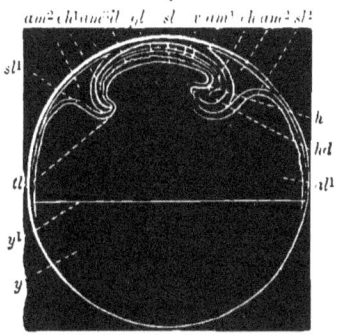

Fig. 208.

Fig. 208. The same as figs. 204 to 207, but further advanced. A profile view. 2 diam. *y*, the yolk; $y^1$, the surface of *y*; $al^1$, the clear, albuminous space; *gl*, the original germinal disc, now, more properly speaking, the body; *hd*, the head; *tl*, the tail; *ch*, the anterior, and $ch^1$, the posterior end of the *chorda dorsalis*; *v*, the primitive vertebræ; $am^1$, $am^3$, the edges of the fold of the amniotic sac; $am^2$, $am^2$, the space beneath $am^1$, $am^3$; *il*, the intestinal layer; *h*, the heart, a channel in the intestinal layer (*il*); *sl*, the subsidiary layer; $sl^1$, $sl^2$, that part of the subsidiary layer in which the vascular area is formed. —*Original*.

forms a nearly closed tube by the folding together of its edges. In the head this tube is wide and deep, (at $hd_{,}$) and there constitutes the anterior end of the nervous cord, *i. e.*, the *brain*. Consentaneously with the formation of the hollow nervous cord, the juxtaposed edges of the halves of the *vertebral layer* become elevated and form a sheath about it. Consequently, we find this layer here, as seen in profile, at the same level with the upper edge of the nervous cord, and extending below to the horizon of the lower face of the *chorda dorsalis* ($ch$, $ch^1$). The extent of this layer is rendered conspicuous, in this profile view, by the presence of a few of the preliminary vertebræ, ($v_{,}$) which have developed from it by a transverse division of its whole thickness into successive squares. The principal additional feature is observable in the inception of the *intestinal layer* ($il$). This is formed by the separation of a stratum of cells from the upper face of the subsidiary layer; first at its central portion ($sl$) along the lower face of the body, and then extending centrifugally, not only as far as the head, tail, and sides, but eventually beyond these points, as we shall see presently. We may also see at this period the manner in which the blood-system begins to form, which is by a mere hollowing of channels in the upper surface of the subsidiary layer. The first distinctly bounded cavity of this system is that of the heart, ($h_{,}$) which, almost from the beginning of the formation of the blood-vessels, indicates its character by feeble pulsations. It is a simple, broad, and short tube which lies lengthwise exactly in the lower mid-line of the body, and a short distance behind the head. At the present period the blood merely surges backward and forward in the channels, under the influence of the contraction and expansion of the heart.

I shall next lay before you the illustration of a phase of development which is considerably in advance of the last one, but in

Fig. 209. The same as figs. 204 to 208, in a more advanced state of development than the last. 2 diam. A bird's-eye view of the embryo. *vs*, the periphery of the yolk ; *y*, the yolk ; $y^1$, the edge of the upper surface of *y ; hd*, the head ; *tl*, the tail; *n*, the region of the nervous cord ; *ch*, the anterior, and $ch^1$, the posterior end of the *chorda dorsalis* ($ch^2$) ; *h*, the heart; *vn*, the median blood-vessel :

which no new organs are introduced whose origin cannot be easily understood without going through all the intermediate stages. In regard to the body proper, two things will be noticed at once: the one is that it has fallen on its side, and the other that it is entirely enclosed by the *amniotic sac* (fig. 209, $am^4$, $am^5$). Another

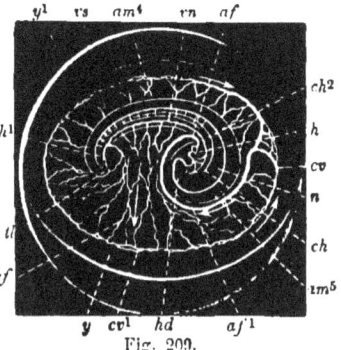

Fig. 209.

prominent general feature lies in the large area of blood-vessels ($ef$, $cv$, $cv^1$, $af^1$). These vessels are but an extension of those whose initiatory formation was pointed out in the last stage. The principal chamber of this system, the *heart*, ($h$,) has become transformed into a one-sided, nearly double cavity, by folding upon itself and narrowing the space between its two halves. The blood-vessels are hollowed in the subsidiary layer in two sets, which are respectively named the *efferent* and *afferent* vessels, *i. e.*, those which carry the blood away from the heart and those which bring it to it again. In the former phase these vessels were so undetermined in their course, and so intermingled, that the blood, as I stated, merely surged backward and forward in the same channels; but since that period they have extended more widely and far beyond the body, and are separated into two groups, which are arranged in this wise: taking the heart ($h$) as a starting-point, the efferent vessels commence with a single current ($vn$) which passes from the heart backward and along the mid-line of the lower face of the body to its end. From this single one the vessels of the great net-work (*vascular area*) arise right and left, and pass, with a few forkings ($ef$) and inosculations, to a considerable distance beyond the body, where they unite in an irregular circular channel, the *vena terminalis*, ($cv$, $cv^1$,) which carries the blood forward beyond the head and

$ef$, the efferent vessels of the vascular area; $cv$, $cv^1$, the circular vessel (*vena terminalis*) of the vascular area; $af$, $af^1$, the afferent vessel; $am^4$, $am^5$, the amniotic sac. — *Original.*

empties it into a large single vessel, the *afferent vessel*, ($af$, $af^1$,) by which it is transported directly to the posterior chamber (*auricle*) of the heart.

By the increased incurvation of the head ($hd$) and tail ($tl$), and the infolding of the sides of the body, the space which lies between these parts is pretty nearly isolated and shaped into a body-cavity. At its anterior region this is entirely closed over, from the head to the point, just behind the heart, where the afferent vessel ($af$, $af^1$) enters. The preliminary vertebræ are developed along nearly the whole length of the body in a backward direction, and, as a natural consequence, the tubular nervous cord, ($n$,) or spinal marrow, has likewise closed over to an equal extent. In front, the vertebræ have extended only a short distance, in fact, as far as they will ever appear as distinct bodies. At this period the eyes and ears are so far advanced in development as to be clearly recognizable in their respective positions.

But one stage further we find the relations of the organs and their state of development very much in the condition of those of some of the more lowly organized fishes, and, in certain respects, like those of the *Lancelet*, (p. 226, fig. 133,) which I described in a previous lecture. This resemblance is especially noticeable

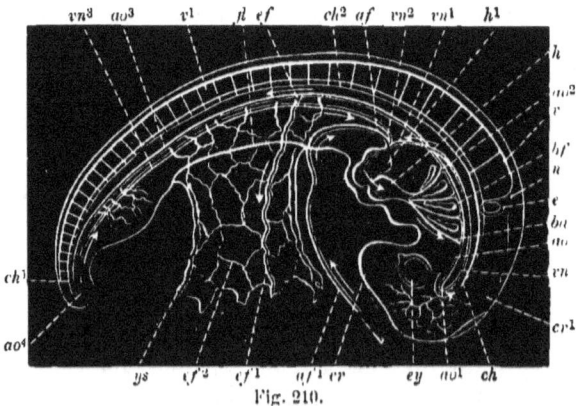

Fig. 210. The same as figs. 204 to 209, but more highly developed. 6 diam. A profile view of the body proper and a part of the vascular area. The amniotic sac has been removed. *ey*, the eye; *e*, the ear; *cr*, the anterior end of the

in the series of transverse fissures (fig. 210, *bf*) on each side of the neck, and the blood-vessels (*ba*) which pass between them, when respectively compared with those of the Lancelet (fig. 133, *br, bo*). On account of this resemblance — in fact by some assumed as an identity both in form and function — the fissures have been called by embryologists the *branchial fissures*, and the vessels the *branchial aortæ*, the former corresponding with the passages between the gills of fishes, and the latter with the vessels which supply the gills with blood. In respect to the whole bloodsystem, there are several additions to, as well as modifications of it, beyond what obtained in the last phase. These will be most easily understood by a description of the courses of the vessels as they now exist. The blood passes from the anterior chamber (*h*) of the heart in a forward direction, and immediately separates into the four branchial aortæ, (*ba*,) which carry it to the main aorta, (*ao, ao*$^2$ *ao*$^3$,) that runs along the upper mid-line of the general cavity of the body. Within the latter the blood passes in two directions. One part of it goes forward into the head in what is called the *cephalic aorta*, (*ao*,) which, after branching there, unites (at *ao*$^1$) its scattered currents into two parallel, recurrent vessels (*vn*) which run on each side of the aorta to a point (at *vn*$^1$) near the posterior cavity (*h*$^1$) of the heart, and there join that organ at the same place with the vessels which come from behind. The other part runs in the main aorta (*ao*$^2$, *ao*$^3$) to the waist of the body, and there again divides into two sets of currents. One of these continues direct to the end of the tail in the main aorta (*ao*$^3$), and

brain; *cr*$^1$, the middle region of the brain; *n*, the posterior end of the brain; *v, v*$^1$, the primitive vertebræ, or *vertebræ dorsales*: *ch*, the anterior end, and *ch*$^1$, the posterior end of the *chorda dorsalis*; *ch*$^2$, the *chorda dorsalis* as seen through the *vertebræ dorsales*; *h*, the anterior, (*ventricle*,) and *h*$^1$, the posterior (*auricle*) chamber of the heart; *ba*, the branchial arteries; *ao*, the cephalic aorta; *ao*$^1$, the end of *ao*; *ao*$^2$, *ao*$^3$, the main aorta; *ao*$^4$, the end of *ao*$^2$. *ao*$^3$, where it joins the afferent vessels (*vn*$^3$); *ef, ef*$^1$, the main efferent vessel; *ef*$^2$, the net-work of efferent vessels; *af, af*$^1$, the afferent vessel, or return current, of the vascular area; *vn, vn*$^1$, the cephalic vein; *vn*$^2$, *vn*$^3$, the abdominal vein, or afferent vessel of the right side of the body; *hf*, the branchial fissures; *fl*, the right edge of the abdominal aperture; *ys*, neck of the nutrient organ. — *Original*.

there branching and reuniting,—as in the head,—passes forward in a single current ($vn^2$, $vn^3$) along each side of the mid-line of the body to the posterior end of the auricle ($h^1$) of the heart. The currents of the other set arise at several points, right and left, along the abdominal aorta, ($ao^3$,) and pass out of the body in numerous channels, ($ef$, $ef^1$, $ef^2$,) of varying calibre, into the great net-work of vessels of the vascular area, and thence into the as yet developing, half-formed vessels in the depths of the yolk mass. From these the blood is gathered into return currents, which finally unite in the great afferent vessel, ($af$, $af^1$,) that passes into the abdominal cavity just behind the heart, and empties its contents into the auricle ($h^1$) at the same point with the termination of other afferent vessels ($vn^1$, $vn^2$). The approximating sides ($fl$) of the body have not only narrowed the lateral extent of the general cavity, but have closed over a considerable space just in front of the tail. In consequence of this, the membrane or layer over which the vessels of the vascular area are spread is narrowed to a sort of neck ($ys$). In view of this we might say that the yolk is contained in a great saccular prolongation of the sides of the abdominal region; and in consideration that the blood-vessels plunge through it in every direction, and divide it into a sponge-like mass, the name *nutrient organ* would not be inappropriate for it; at least, it will serve to indicate its nature and peculiar function at this period of life.

There is no conspicuous change in the development of that part of the nervous cord which lies behind the head, but that portion which lies in front of the ear, ($e$,) and corresponds to the brain, has commenced the division of its cavity into two chambers, by slight constrictions. The first division extends from the end ($cr$) of the head backwards to a point nearly opposite the end ($ch$) of the *chorda dorsalis*, and corresponds to the cerebral region of the adult (see p. 230, fig. 134); the second division reaches from the termination of the last to the ear, ($e$,) and answers to two parts of the brain; namely, in front, to the posterior lobes of the *cerebrum*, and behind, ($n$,) in the neighborhood of the ear, to the *cerebellum*, or posterior region of the brain (see

p. 230, fig. 134, *cr*). The latter (*n*) is not as yet closed over so as to form a perfect tube; in fact, even in the adult its opposite edges are united by a thin membrane which is rather the sheath of, than a part of, the tube itself. The eye (*ey*) is very large in comparison to the size of the head, and, as its position indicates, originates in connection with the anterior part of the brain, whereas the ear (*e*) is developed from a totally different region, *i. e.*, the cerebellum. Both the eye and ear are saccular prolongations or projections from the regions whence they originate; the one being formed into a cup-shaped body into whose mouth the rays of light pour from every direction, and the other taking the guise of a drum, against whose sides the reverberations of the air pulsate with varying intensity.

Beyond this stage the process of development has more especial reference to the formation of a particular kind of reptile, *i. e.*, a turtle, than to the construction of a Vertebrate in general; and I shall therefore merely indicate the further growth of the embryo in as brief terms as possible. The stomach and intestine are formed by a folding together of the right and left halves of the membrane in which the blood-vessels run, in such a way as to leave a hollow channel extending from one end of the body to the other, except at that part of it where the yolk sac is attached, and there it lies in open communication with the latter. The lungs arise as hollow, saccular prolongations of the sides of the throat. The liver is a thickening of the same membrane from which the stomach was formed, and is perforated in every direction by a dense net-work of blood-vessels. The reproductive organs arise in intimate connection with the posterior end of the intestine. The legs develop as direct outgrowths from the right and left sides; first as mere rounded protuberances, and finally expanding into flattened pads or paws. The shield is formed by a lateral expansion of the regions of the back and front, with anterior and posterior apertures into which the head and legs can be withdrawn. During the progress of the abovementioned developments, and the increasing size of the body proper, the *nutrient organ* becomes gradually lessened, and finally, just after the hatching of the young, is completely drawn into the body, and resorbed by the intestine.

### § Conclusion.

Thus it appears that there is a plainly visible, intelligent, *controlling power*, which is manifested, with *unvarying regularity of character*, in each of the five groups of animals. Shall we now undertake to say how far, how minutely, that power operates; and shall we assume that we can tell at what point in time or place this power ceases to act *undisturbed* in the regularity of its proceedings, and allows itself to be swerved from its course by the apparently disturbing influence of circumstances?

Who shall say that these circumstances have not been provided for, or even regulated in the succession of events so as to become a part of the plan?

We all know that the *physical agents*, light, heat, electricity, magnetism, &c., have a *law* according to which they evince their operations, — as the science of meteorology teaches; and if, now, these so-called secondary causes have a method among themselves, we should expect to find them likewise affecting the processes of *organic* nature in the same orderly manner that they affect each other interchangeably; and consequently evincing the presence of the same directing hand in the one case as the other.

We cannot arbitrarily assume that these forces are included in the plan, in one case, — for instance, *heat* and air in hatching the bird's egg, — and reject them in another, or class them as *accidental*.

How far time and the progress of science may lead us on to a better understanding of the mode of operation of these forces, it is impossible to foresee; but we may, I think, venture the conjecture, that, since in all the thus-far-known phenomena of Nature we have learned to recognize a more or less intimate and direct relation to each other, either in the condition of *influencing* or of *being influenced*, we shall presently discover that many of the so-called variable influences and accidents have a true periodicity of *intensification* and *relaxation*, which hitherto has seemed to be irregular, simply because of the individual character of the organism upon which they operate.

Every day reveals to us new channels in the courses of nature; but as we trace them back to their source, we find them all to be the branches of *one great current*, which forces everything before it onward and straightforward into the universal ocean, — *the end of all things* and the beginning of the new; that great reservoir from which the elements of all beings are derived, and to which they all return, in one eternal circle of changes, from the elaborate composition of the body of the growing man to his going down again into the disintegrating, fluttering atoms, and their final diffusion into the primitive vapors.

## ADDENDUM.

ON page 162, line 20, at "*vibrating lash*," add as a note the following: — I have ascertained by observation, while these pages were going through the press, that *this so-called lash is an optical illusion*, and that it is really a row of closely set, vibrating cilia seen edgewise, or *foreshortened*; just as the teeth of a comb would appear if their points projected toward the eye. This has been confirmed by investigations of other species of Vorticellidæ, e. g. *Epistylis galea*, Ehr?, *E. grandis*, Ehr?, *Vorticella nebulifera*, Ehr., *Carchesium polypinum*, Ehr., and *Trichodina pediculus*, Ehr.; and I conclude, therefore, that the so-called "*bristle of Lachman*," described and figured as long ago as 1856, (Lachman in "*Müll. Archiv.* 1856,") does not exist in nature. See, for further details, my forthcoming memoir on the "*Anatomy and Physiology of Trichodina*," in the "Memoirs of the Boston Society of Natural History."

CAMBRIDGE, Mass., *November* 16, 1865.

# INDEX.

## A.

Abdominal pore, 228.
Acanthometræ, 50.
Actinophrys, 44, 156.
Adults, the primordial state of animals, 106.
Aërification, 211.
Affinity, chemical, 7.
    inorganic, 7.
    organic, 7.
    vital, 7.
Agassiz on the systematic position of Infusoria, 174.
    the parasites of Hydra, 174.
    Vorticellidæ and Bryozoa, 174.
Albertin, 274.
Albumen and yolk of eggs, diverse origin of, 35.
    of the egg, 33.
Alcohol, composition of, 8.
Algæ, 134, 155.
Alternate generation, 67.
Ammocœtes, 125.
Ammonia, composition of, 8.
Amniotic sac, 306, 309.
Amœba, 9.
    organic functions of, 11.
Amphioxus, 126, 226.
Anchorella, 272, 274.
Animal-egg, the, 88.
Animal-flower, 57.
Animals, the lowest forms known, 15.
    allied by progressive relations, 5.
    false ideas of development of, 6.
    and plants, distinction, 131, 151.
Animate being, the universe ruled by an, 5.
Aquiferous circulatory system of Echinoderms, 186, 190.
Arachnida, 220.
Archeopteryx, 277.
Area, vascular, 309.
Arrangement, spiral, in corals, 60.
Articulata, 119, 214.
    and Protozoa, relations of, 241.
    compound nature of, 214.
    and Zoöphyta, 246.
    development of, 208.

Artificial division, experiments on Planaria, 92.
    of Amœba, 95.
    of Sea-anemones, 89.
    reproduction after, 89.
Ascaris, 36.
Asteracanthion, 181.
Aurelia, 67, 70.
    motions in decomposing cells, 98.
Axial plane, of Mollusca, 197.
Axolotl, 263.

## B.

Bacterium, 18, 23, 102.
Balbiani on eggs of Infusoria, 31.
Beetle, 221.
Beings, all living, chemical composition of, 6, 8.
Bell-animalcules, 161.
Bilaterality, 128, 267.
    the predominant feature in animals, 249.
    early appearance of, in animals, 251.
    relation of, to reproduction, 266.
    of Zoöphyta, 179.
Bipolarity, 267.
Birds' eggs, 37.
Bisymmetry and bilaterality, 185.
Blanchard on Planarians, 253.
Blood circulation of Echinoderms, 186, 191.
Bonellia, 215, 246.
Brain, experiments on removal of, 127.
    duplicity of, 127.
    relative position of, among animals, 125.
Branchial aortæ of the embryo, 311.
    fissures of the embryo, 311.
    ganglion, 202.
Bryopsis, 139.
Bryozoa, 195.
    and Zoöphyta, 246.
Buccal plates, 190.
Budding, individuals arising by, 54.
    of Bryozoa, 244.
    Fredericella, 244.
    Sea-anemone, 60.

## C.

Caddice-fly, development of, 298.
Carrion-beetle, 221.
Caudina, 187, 246.
Cells, the egg the type of all, 33.
Cellular structure of sponges, 42.
Centimetre, 17.
Centrum, 231.
Cephalization, 211, 220, 230.
    Dana on, 220.
    etc., note on, 193.
Cephalopoda, 207.
    position of shell of, 212.
Cephalothorax, 219.
Ceratium, 147, 156.
Cerebral ganglion, 202.
Cereus, 58, 179.
Chelydra serpentina, 304.
Chelyosoma, 253.
Chemical affinity, natural, 7.
Chilodon, 167.
Chipping sparrows, 277.
Chlamidomonas, 133, 145, 156.
Chorda, 143, 155.
*Chorda dorsalis*, 229, 305.
Ciliata, 286.
Ciliated infusoria, 148.
Circulation in Insects, 224.
    plants, 166.
Cladophora, 140.
Claparède on Opalina, 245.
Classes of animals, transitions among, 254.
Classification of Cuvier, 117.
    Lamarck, 117.
    of the principal groups, 255.
Clava, 75.
Cobbold, on the development of a Sea-anemone, 289.
Collar, nervous, 125, 225.
Compound individuals, 54.
Confervæ, spores of, 134.
Consanguinity, relations of, 236.
Contractile vesicle, 151, 162, 168.
    of Actinophrys, 47.
Contractility in animals, 141.
Cornuspira, 13.
Coryne, 75, 77, 242.
Creation, primary and continued, 30.
Creator, primary interposition of, 4.
Crustacea, 219.
    cephalization in, Dana on, 220.
Currents of circulation in sponges, 43.
Cuttle-fish, 207.
Cuvier, classification of, 117.
Cyclops, 220, 268, 274.
Cytoblastema, 45.

## D.

Dana, J. D., on cephalization, 220.

Danielssen and Koren on Siphonactinia, 180.
Decomposition, experiments on, 95, 97.
Degrees, phenomena of nature differ by, 113.
Demodex, 274.
Dendrocœlum, 92, 268.
Design, thoughtful, evidence of, 5.
Development of animals, 283.
    Articulata, 298.
    Caddice-fly, 298.
    Mollusca, 294.
    Mystacides, 298.
    Protozoa, 285.
    Sea-anemone, 178, 289.
    Vertebrata, 303.
    Zoöphyta, 289.
    old theory of, 158.
    progressive, 160.
    phases of, common to all animals, 158.
    theory, old, 3.
    new, 3.
Difflugia, 11.
Digestion in Rhizopoda, 14.
    mode of, among tape-worms, etc., 100.
Digestive system of Infusoria, 167.
Diptera, 222.
Disc, germinal, 305.
Duality of individuals, 267.
Duck-billed quadruped, 53.
Duck-mole, 279.
    young of, 53.
Dulse, 143.
Dysteria, 171.

## E.

Echinodermata, development of, 290.
Echinoderms, 181.
Egg, mulberry state of, 285.
    of Infusoria, 31.
    Laomedea, 33.
    Rabbit, 32.
    Sow, 32.
    place of origin of, variable, 52.
    segmentation of, 284.
    structure of, 31, 33, 36.
    the, and the animal identical, 52.
        is the first phase of the animal, 52.
Eggs, bipolar bodies, 40.
    floating in the air, 29.
Ehrenberg on Infusoria, 167.
Embryo, degree of dependence upon the parent, 53.
    relation of, to external causes, 53.
    to the parent, 53.
Embryonic disc, 304.
    eggs, 40.
    forms, Von Baer on, 252.
    state of animals, 40, 252.

## INDEX.                                                                          319

Encrinite, fossil, 128.
Eolidiceros, 253.
Epeira, 221.
Ephyra of Aurelia, 68.
         Pelagia, 75.
Epistylis, 161, 237.
Euglena, 134, 144, 156.
Eye-animalcule, 144.
Eye-spot of animalculæ, 144.
    false, in Zoöspores, 140, 144.

### F.

Fallopian tube, 234.
False circulation in Actinophrys, 47.
Family traits, 276.
Feathered reptile-like bird, 277.
Ferment cells, 20.
Firolidæ, 253.
Fissigemmation explained, 81.
Flat-worm, 92.
Fly, 222.
Flying lizard, 278.
Fowl's egg, structure of, 37.
Fredericella, 198, 243, 247.
        budding, 244.
Funnel of Cuttle-fish, 208.

### G.

Garden spider, 220.
Generation, spontaneous, 15, 110.
Geology in past times, 104.
Germigenous organ, 36.
Germinal disc, 305.
    dot, 35.
    layer, 304.
    vesicle, 32, 35.
        character of, 33.
    layer of Articulate embryo, 299.
Globe-animalcule, 153.
Glosso-pharyngeal nerve, 231.
Gonium, 155.
Grand divisions of the animal kingdom, 160.
Green Dulse, 143.
Groups of the animal kingdom, 121, 160.
Growth, parallelisms of, 159.
Gryphæa, 203.

### H.

Hartshorn, composition of, 8.
Hawk-moth, privet, 223.
Head-chest, 219.
Helix, 203.
Hemicosmites, 128, 253.
    bilaterality in, 128.
Hereditary transmission, 279.
    influence, 279.
Heteromastix, 134, 146, 156.
Heteromita, 134.

Holothuria, development of, 291.
    bilaterality of young, 249.
Hood of Nautilus, 213.
House-fly, 222.
Hydra, 55, 84, 89, 177, 237.
    artificial division of, 89.
    budding, 56.
    experiments on, 89.
    fresh-water, 55.
    reproduction of lost parts, 89.
Hydro-medusæ, 73.
    individuality of, 73.

### I.

Ideal eggs, 34.
    relations of the five types, 236.
    transitions, 236.
    types, 122, 126.
Individual, the true, 54.
Individuality, 54, 75, 91.
Infusoria, ciliated, 148.
    development of, Agassiz on, 174.
    development of, in boiled solutions, 15, 101.
    resistance to heat, 26.
Inorganic affinity, 7.
    bodies, 8.
Inosculating groups, 276.
Insects, 221.
    formation of eggs in, 35.
Intelligence of Infusoria, 170.
Interchangeable relations of animals, 268
Interlocked groups, 276.
Intestinal layer, 308.

### J.

Janus, 253.
Jaw-like bodies of Infusoria, 173.
Jelly-fish, 177.

### K.

Kolpoda, 18, 102.

### L.

Lamarck, classification of, 117.
Laminaria, 143, 155.
Lancelet, 126, 226.
Laomedea, egg of, 83.
    egg-segmentation of, 284.
Larva of Caddice-fly, 302.
    Mystacides, 302.
Laver, 143.
Law of cephalization, 220.
    reproduction by budding, 109.
Layer, germinal, 304.
    intestinal, 308.
    subsidiary, 304.
Lepidosiren, 263.
Lereboullet on monstrosities, 85, 267.

Lereboullet on development of Lymneus, 294.
Mollusca, 294.
Lerneans, 274.
Life, limits of manifestation of, 95.
   lowest condition of, in the egg, 28, 31.
   origin of, 3.
   organic, 8.
   principle of, 7.
Linguatula, 268, 272.
Lithocampe, 49, 156.
Loligopsis, 206.
Longitudinal axis of Sea-anemone, 179.
   Starfishes, 183.
   repetition of parts in Articulata, 215.
Lowest kinds of animals, 9.
Lydella, 274.
Lymneus, development of, 294.
   native, young of, 297.

## M.

Madreporic canal, 185, 191.
   in young Holothurian, 293.
Madreporiform body, 185, 191.
Mammalia, 230.
Mammary gland, 235.
Mantle of Mollusca, 201, 208.
Medulla oblongata, 234.
Medusa of Coryne, 77.
Medusæ developed directly from the egg, 75.
Meduso-genitalia, 75.
Menobranchus, 263.
Method in the operations of the Creator, 103.
Metridium, 57, 84, 178, 247.
Milne Edwards on self-division of worms, 81.
Mimetic forms, 248.
Mites, 274.
Mollusca, 118, 195.
   development of, 294.
   and Protozoa, relations of, 243.
Monads, 18.
Monomerous, the type, 195.
Monostroma, 155.
Monstrosities, 86.
Motile decomposing cells, 98.
Motory nerves, 234.
Mould, blue, 21.
   white, 21.
Mulberry state of the egg, 285.
Müller on self-division of worms, 81.
Multiple self-division, 80.
Muscle, decomposition of, 95.
Myrianida, 80, 215.
Mystacides, development of, 298.
Myxine, 125, 253.
Myzostoma, 275.

## N.

"Natural Theology" of Paley, 104.
Nautilus, 213.
   position of shell of, 212.
Necrophorus, 221.
Nervous collar, 125, 225.
   system, correspondence of, in the grand divisions, 125.
   system of Stentor, 64.
Note on scientific property, 37, 304.
Notochord, 229.
Nucleolus, 32.
Nucleus, 20.
Nutrient organ of the Vertebrate embryo, 312.

## O.

Oblique or spiral type, 160.
Omne vivum ex ovo, 85.
One-celled animals, 99.
Opalina, 174, 245.
Order of things, 264.
Organic affinity, 7.
   and inorganic bodies, difference between, 48.
   life, 8.
Ornithorhynchus, 53, 279.
Oscillatoria, 25.
Ostrea, 200.
Ovary, 225.
Oviduct, 225.
Owen on Amphioxus, 229.
   Archeopteryx, 278.
   secondary causes, 102.
Oyster, 199.

## P.

"Paley's Philosophy," 5
Paramecium, 163.
   development of, 286.
Parasites of Hydra, Agassiz on, 174.
Parentage, common, of species, 277.
Pectinatella, 196, 243.
Pelagia, 75.
Pentatremites, 183.
Physical agency in the development of animals, 53.
   forces, 3.
   force, relation of the egg to, 100, 102.
   laws, 3.
Physicists, older, laws of, 5.
Phytozoa, 156.
Planaria, artificial division of, 92.
Planariæ, 275.
Planarians, articulate character of, 253.
Plane of the axis of Zoöphyta, 243.
   Protozoa, 243.
Plan of creation, original, 106.
Plant-animals, 156.
Planula of Aurelia, 242.
Pleuronema, 148, 156, 170.

## INDEX.                                321

Pneumodermon, 252.
Podophrya, 51, 157, 174, 237.
    development of, 286.
Polarity in animals, 127.
Poles of the egg, 33, 35.
Polycystinæ, 49.
Polymerous, dorso-ventrally, the Zoophyta, 195.
    uro-cephally, the Articulata, 195.
Polypi, development of, 289.
    anatomy of, 57, 177.
Pore, abdominal, 228.
Primarily created animals, condition of, 105.
Primitive stripe, 305.
Primordial state of the first created animals, 105.
Privet Hawk-Moth, 222.
Progressive steps, animals related by, 265.
Property, scientific, note on, 37, 304.
Prostomium, 35.
Protean animalcule, 9.
Protococcus, 133, 136.
Protozoa, 160.
    and Articulata, relations of, 244.
    Zoöphyta, relations of, 237.
    Mollusca, relations of, 243.
    bilaterality of, 175.
    development of, 285.
    progressive development of, 40.
    self-division of, 61.
Protula, 251.
Pseudopodia, 12, 44.
    rapid movement of, 13.
Psolus, 192.
Pterodactylus, 278.
Purkinjean vesicle, 32.

### Q.

Quatrefages on Amphioxus, 228.

### R.

Radiata, bilaterality in, 128.
Rank, relative, of Articulata, 124, 214.
    Mollusca, 124, 214.
Red-snow plant, 135.
Reduction of repeated parts in Articulata, 215.
    Zoöphyta, 187, 192.
Relation of organs in typical forms, 126, 159.
Reproduction by budding, law of, 109.
    of lost parts in animals, 94.
    relation of, to form, 265.
    bilaterality, 266.
Reptilian fishes, 263.
Respiratory branches, 190.
Rhizogeton, 73, 242.
Rhizopoda, 14.

Right and left in animals, 127.
Rotalia, 14.
Rotifera, 275.
Round-worm, formation of egg in, 36.

### S.

Sagitta, 96.
Saltatory cilia of Infusoria, 150.
Sand-eel, 227.
Saprolegna, 133, 141.
Sarcode, 11.
Scientific property, note on, 37, 304.
Scorpions, 220.
Scyphostoma of Aurelia, 67.
    tentacles of, 242.
    organization of, 241.
Sea-anemone, 57, 177.
    young of, 177.
    development of, 289.
Secondary causes, Owen on, 102.
    relation of embryo to, 111.
    prevalence of, a matter of degree, 112.
Seeds floating in the air, 29.
Segmentation of the egg, 284.
Self-division among Corals, 61.
    Protozoa, 61.
    Vertebrates, 85.
    in Tape-worms, 82.
    of Myrianida, 80.
    worms, 80.
    Sea-anemone, 61.
Shell, relative position of, in Cephalopoda, 212.
Shrimps, 274.
Siamese twins, 86.
Similarity of all animals at earliest period of growth, 111.
Siphonactinia, 180.
Siphydora, 41.
Sipunculacea, 250.
Siren, 263.
Snail, 203.
    development of, 294.
Snapping turtle, 304.
Song-sparrows, 277.
Species, affinities among, 276.
Sphinx, 223.
Spiders, 220.
Spinal column, rudiment of, 229.
    nature of, 231.
Spinal cord, double, 127.
Spiracles, 224.
Spiral type of animals, 160, 175.
Spirillum, 23.
Spirostomum, egg of, 31.
Sponge, 40, 156.
Spontaneous generation, 15.
    origin of theory, 28.
    probability of, 106.

Spores, 136.
  of plants, resistance to heat, 23.
Squamella, 268.
Squid, 206.
Starfish, 177, 181.
  reproduction of lost parts, 91.
Stentor, 62, 84, 160, 163.
  budding of, 65.
  development of, 288.
  egg of, 31.
  nervous system of, 64.
  young of, 148, 288.
Sting-blubber, 70.
Stomachs of Infusoria, 167.
Stripe, primitive, 305.
Subordination of types to bilaterality, 128.
Subsidiary layer, 304.
Suctoria, 285, 286.
Sunfish, 67.
Symbolical form of animals, 122, 158.

T.

Tænia, 83.
Tape-worms, 82, 215.
Tardigrada, 268, 274.
Tentacles, 57, 177.
Tetraspora, 155.
Thorax of Insects, 222.
Thuret on spores of aquatic plants, 133.
  sea-weeds, 133.
Torula, 20.
Tracheæ, 224.
Transitions among classes, 254.
  minor groups, 262.
  species, 276.
  ideal, 236.
Trembley's experiments on Hydra, 89.
Trepang, 120, 177, 187.
Trumpet-animalcule, 62.
Tubularia, 75, 79, 238.
  bilaterality of, 239.
  arrangement of tentacles of, 239.
  development of tentacles, 239.
Turtle, snapping, 304.
  development of, 304.
Turtles and Tortoises, development of, 303.
Turtles' eggs, 37.
Twins, 86.
Type of division, 160.
Types of animals, 117, 122.
  the five great, 122, 160, 248.
Typical animal, 265.
  bird, 277.
  egg, 32.
  forms, relation of organs in, 126.

U.

Ulva, 143, 155.
Uniformity of the organization of Mollusca, 195.

Up and down, relations of animals to, 121.

V.

Van Beneden on Linguatula, 273.
  Lerneans, &c., 274.
Van der Hœven, on Vorticellæ, 243.
Vascular area, 309.
Vaucher on spores of plants, 132.
*Vena terminalis*, 309.
*Vertebræ dorsales*, 311.
Vertebræ, preliminary, of embryo, 310
Vertebral arches, 232.
  column, elements of, 125.
  layer, 307.
Vertebrata, 118, 226.
  development of, 303.
Vibrio, 96.
  baccillus, 25.
  rugula, 18, 102.
  resistance to heat, 23, 26.
Vibrios, false, 97.
Vital affinity, 7.
Vitality, degrees of, in eggs, &c., 99.
Vitelligenous organ, 36.
Volvox, 153, 150.
Von Baer on embryonic types, 252.
Vorticellæ, 161, 237.
Vorticellidæ, Agassiz on, and Bryozoa, 174.

W.

Wagnerian vesicle, 32, 35.
Water, composition of, 8.
What is an animal, 132.
Winged mammals, 278.
  quadrupeds, 278.
Winter-eggs, 275.
Worm-shaped embryo of Starfish, 248, 250.
  Molluscan, 248.
Wright, Dr. Strethill, on Zoöteira, 50.
Wyman, Prof. J., experiments of, 15.

Y.

Yeast-plant, 20.
  development of, 20.
Yolk and albumen of eggs, diverse origin of, 35.
  of the egg, 33.

Z.

Zoöglœa, 19.
Zoöphyta, 120, 177.
  development of, 289.
  and Articulata, 246.
  Bryozoa, 246.
  Protozoa, relations of, 237.
Zoöspores, 140.
Zoöteira, 50, 156.
Zoöthamnium, 163, 175.

www.ingramcontent.com/pod-product-compliance
Lightning Source LLC
Chambersburg PA
CBHW021202230426
43667CB00006B/515